SURF SCIENCE

AN INTRODUCTION TO
WAVES FOR SURFING

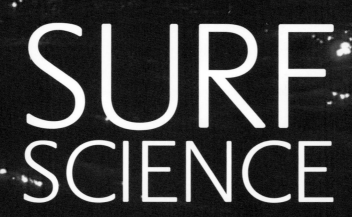

SURF
SCIENCE

AN INTRODUCTION TO
WAVES FOR SURFING

Tony Butt &
Paul Russell

with Rick Grigg

University of Hawaii Press
Honolulu

Alison Hodge

First published in 2002 by
Alison Hodge, Bosulval, Newmill, Penzance, Cornwall TR20 8XA, UK
www.alison-hodge.co.uk   info@alison-hodge.co.uk
Reprinted in 2003

This edition is published in the UK by Alison Hodge, and in the USA by
University of Hawaii Press, 2840 Kolowalu Street, Honolulu, HI 96822,
uhpbooks@hawaii.edu

ISBN 0 906720 36 2 2nd revised edition (UK)
ISBN 0 8248 2891 7 (USA)
(ISBN 0 906720 31 1 1st edition)

British Library Cataloguing-in-Publication Data
A catalogue record for this book is available from the British Library.

Library of Congress Cataloging-in-Publication Data
A catalog record of this book is available from the Library of Congress.

Edited, designed and originated by
BDP – Book Development and Production, Penzance, Cornwall, UK
Cover photograph by Robert Gilley
Cover design by Christopher Laughton

Printed and bound in Singapore

CONTENTS

PREFACE

The idea for this book came from a series of articles on oceanography and surfing written by Tony Butt for *The Surfer's Path* magazine. We aim to fill the gap between surfing books and waves books. *Surf Science* is the first book to talk in depth about the science of waves from a surfing point of view. Most books on the market are either general surfing books with a fairly poor oceanography section, or highly mathematical textbooks, totally beyond most people. This is because waves are very difficult to explain, so writers tend either to go crazy with the equations, or to provide scanty, inaccurate descriptions.

We have made every effort to steer between these two extremes, and to make the subject of wave motion as painless as possible. We use simple, light-hearted analogies to allow readers to visualize some of the more esoteric concepts. For example, the idea of radial dispersion becomes easier to understand if the waves are likened to marathon runners.

Surf Science is written for surfers who, like us, are not content just to go out and surf, but who are fascinated by where the waves come from; what makes every wave different, and what factors affect the behaviour of a surfing break. You do not need a scientific background to read this book. All you need is curiosity and a fascination for waves.

However, we have included some technical material, which enables the book to double-up as a basic text for first-year undergraduates in oceanography and surf science. This extra material is not essential for the understanding of any of the concepts in the book, and it may be completely ignored if desired, with no loss of 'flow' in the text. There is nothing more frustrating than having to wade through a nightmare of incomprehensible scientific jargon which, only after you have painstakingly found out what it all means, turns out to be irrelevant!

These extra concepts are presented in the form of boxes, aside from the main text. The boxes contain some mathematical equations, but these are shown in highly simplified form. If you wish to investigate the more complex and complete forms of these equations, with the proper symbols and definitions, you can look them up in the references given in each box. Equations are some of the worst things for frightening off potential readers. Stephen Hawking, in his famous book, *A Brief History of Time*, writes, 'Someone told me that each equation I included in the book would halve the sales.'

There are also a few terms in the main text with which you may not be familiar. These appear in bold print, and are defined concisely in the Glossary-Index at the back of the book.

The first eight chapters of *Surf Science* describe the life of a surfing wave (or, more precisely, a packet of energy briefly manifest in the form of waves) from before its birth in an oceanic storm, to its final dissipation on the shore. Each of these chapters describes a different process and may be read on its own, or in sequence. Each of the other six chapters is a self-contained description of a useful scientific aspect of surfing.

Each chapter has a number of simple colour diagrams, designed to provide a clear, concise illustration of each concept. The diagrams are complemented by photographs, which provide real examples of some of the ideas, such as convex refraction (defocusing or divergence) and concave refraction (focusing or convergence).

Although many of the examples of surf spots are in Europe, where this book has been written, we have included examples from around the world, so it will be relevant everywhere. It is very important not to forget, for example, that weather systems in the Southern hemisphere revolve the opposite way from those in the Northern hemisphere.

We gratefully acknowledge the help of the following people in the writing of this book: Alex Dick-Read, Joserra Uriarte, Nacho Susaeta, Steve England and Kylie Russell.

Tony Butt and Paul Russell
May 2002

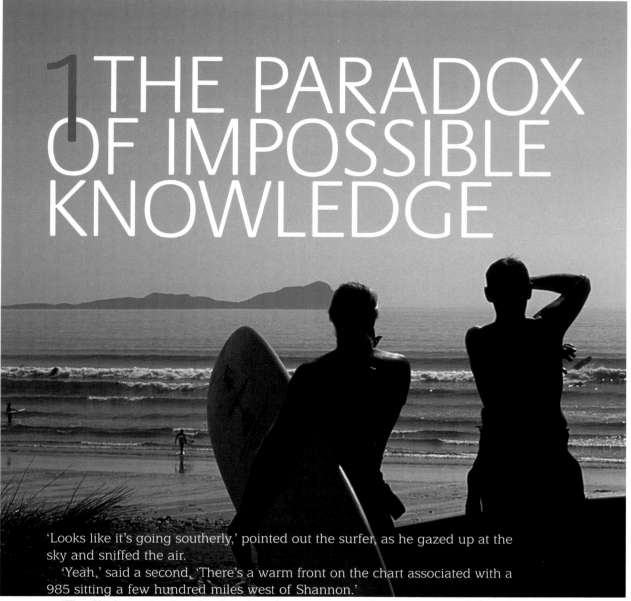

1 THE PARADOX OF IMPOSSIBLE KNOWLEDGE

'Looks like it's going southerly,' pointed out the surfer, as he gazed up at the sky and sniffed the air.

'Yeah,' said a second, 'There's a warm front on the chart associated with a 985 sitting a few hundred miles west of Shannon.'

'The K2 buoy is already showing three metres at fourteen seconds,' said the first. 'I reckon it'll pick up on the push of the tide – what is it anyway, springs or neaps?'

'It's a five point eight,' said the second, after consulting a strange list of numbers screwed up in his back pocket.

To the innocent bystander, this might sound more like scientific terminology than the jargon associated with some sport or leisure activity. Without realizing it, most surfers are also scientists. For without any special effort, a surfer who spends many years waiting for, and riding, different waves in different parts of the world becomes a meteorologist, oceanographer, geographer, linguist and cultural expert. Through his obsession with tapping the ocean's energy to propel him and his board along for a few seconds, the surfer ends up acquiring a large amount of peripheral information. All that waiting, watching, discussing, waiting and thinking gives surfers an insatiable thirst for knowledge, which is typical of scientists rather than sportsmen. This unique facet of surfing gives it a richness rarely found in other activities.

No doubt most of us have asked ourselves – while gazing out at the ocean, or sitting at some surf spot waiting for the next set – questions like these:

- Why is every wave different?
- Why are some waves more powerful than others?
- Why do some peel nicely and some just close out?
- Why, some days, do the waves come in sets of six, and other days in sets of three?
- How would this place work on a north, or indeed a south, swell?
- Why didn't that low produce any surf?
- Where did that swell come from?
- What are waves anyway?

And doubtless some of us have asked ourselves many more obscure questions. Some are readily answerable; others take some thinking about. Yet others – a surprisingly large number – are much more difficult or impossible to answer, even for top oceanographic researchers.

While most surfers are scientists, few scientists are surfers. There are some concepts that surfers know intimately, in a qualitative way, that are barely acknowledged by the scientific community. Usually this is due to a lack of demand for practical application. For example, the **groupiness** of a swell may be studied by engineers for the design of coastal structures. But exhaustive details of the length of time between sets; the number of waves in a set; whether each set has the same number of waves in it; or how the wave heights are distributed throughout the set, are not required for such a study. For surfing, these details are highly relevant. It makes all the difference on a big day if the sets are regularly and widely spaced, as opposed to random, mixed up and close together.

Yet the answers to some questions, that surfers constantly ponder, have been available in the scientific literature for years. A little knowledge about **wave periods**, for example, can explain the difference between a strong, powerful new swell and a weak, gutless old one.

Apart from simply wondering why the waves behave as they do, we probably spend a lot more of our time wondering what the surf will be like in the near future. This has a direct practical application. Since we plan our lives around the surf, being able to predict what the waves will be like tomorrow or next weekend might save us a lot of time and effort. Nowadays, with webcams, surf calls, faxes and surf-forecasting web pages, surfers have access to many efficient ways of predicting the surf, or at least knowing the conditions before setting off on a journey to the beach. Using these facilities is becoming easier and easier, but without a rudimentary knowledge of meteorological and oceanographic terminology, and the meaning of all those lines and symbols on the charts, we run the risk of restricting ourselves to the most basic verbal wave forecasts and reports.

Even the 'non-competitive', 'soul' surfer may be more competitive than he admits. If he can sneak off and score perfect waves alone, or with two of his buddies, while the rest of the surfing world is running around not knowing which beach to head for, then he will. In surfing, the concept of one-upmanship is very much alive. The serious surfer is therefore obliged to know where to go for the latest and best prediction, in order to stay ahead of his competitors. For him, it is essential to know as much as possible about weather charts, wave models, wavebuoys and shipping bulletins, if he is to catch the swell at its best and continue to surf uncrowded conditions.

Within the world of wave and weather predictions lies a strange paradox. Knowing exactly what it is going to be like every day might detract from the interest. Perhaps we need to be slightly in the dark to make life exciting. We need that degree of anticipation, that uncertainty, so that when the perfect day does arrive, we appreciate it. Safe in the knowledge that we will never be able to predict the surf with perfect accuracy, we continue trying. In the process, we learn more and more about how waves work; what affects their quality and quantity, and what affects our ability to enjoy them.

The first part of this book is about a journey. If we want to know what a wave is and how it works, we need to know a little about its history. But the traveller in our journey is not an individual wave, it is a 'packet' of energy.

The journey starts with energy from the Sun entering our atmosphere. This energy drives our weather and wind; generates waves on the ocean; directs these waves towards our coasts, while moulding them into different shapes and forms, and finally uses them to shape the coastline itself. The energy jumps from one form to another, and for a brief moment is manifest in a sloping lump of moving water that we can ride on a board.

Once the Sun's energy enters the atmosphere, the atmosphere is set in motion by the uneven heating of the poles and the Equator, and that strange but vital ingredient, the **Coriolis force**. The motion of the atmosphere is highly complicated and three-dimensional, with a large number of different routes for our energy packet to take. One useful thing this motion gives us is the formation of large, swirling vortices of surface air called depressions. This air rubs along the surface of the water, transferring the energy from one fluid to the other, and generating the first recognizable form of surfing waves, the **windsea**. How this moving surface air ruffles up the surface of the water is still not fully understood by physicists, but we do know that once a wind has been blowing over the ocean for some time, waves are made.

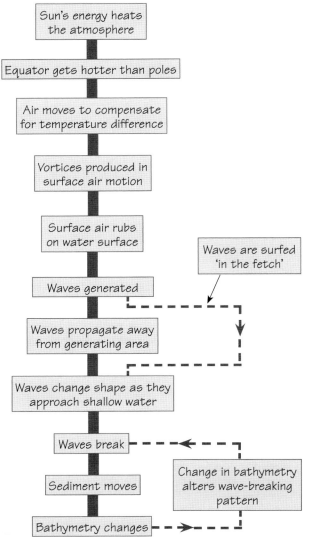

Figure 1.1
Flowchart
summarizing
the different
stages in the
journey of an
energy 'packet'.

As the energy leaves the area of wave generation, it travels within the water, in the form of swell. The swell crawls across the ocean surface, gradually organizing itself into neat groups of undulations heading towards the coast. Once near the coast, these undulations are influenced by the topography of the sea-floor, the **bathymetry**, which bends and warps them into different shapes and sizes. Inevitably, the energy carried by these undulations transforms itself into moving water; in other words, the waves break. It is here that surfers can bleed off an infinitesimal part of this energy and use it for their own entertainment.

In the case of sandy beaches, after the waves have broken, the energy in the moving water is transferred to another fluid – the coastal sediment. The waves move the sediment from one place to another, which in turn alters the way the waves break. At this point, some of the energy is caught in a feedback loop between sediment and breaking waves. Some of it escapes and continues on an endless journey. Figure 1.1 shows the journey of an energy packet. Note that the last part of the flowchart is not relevant to reefs.

The first part of this book is intended to be read as a sequence. The second is a collection of self-contained chapters, each describing a very important scientific aspect of surfing.

An increasing number of surfers live in places where there is just not enough room for the waves they ride to have propagated thousands of miles from some distant storm. Here, with no propagation or dispersion, the journey described in the first part of this book is short-circuited. The waves are ridden in the storm-centre itself. Chapter 9 describes the peculiarities of surfing in the **fetch**, and reminds us that not everybody has the luxury of clean groundswells and ruler-edged lines.

Sometimes, waves generated locally are not wanted. These may ruin a nice clean swell by adding unwanted, short, choppy waves. The **sea breeze**

– a local onshore wind that blows at certain times of the day in warm climates – can, in the space of half an hour, absolutely devastate perfect surf. Chapter 10 explains a few things about the sea breeze: how it works, where it is felt, and when.

The sea breeze depends on temperature differences between the land and the sea. In some parts of the world, thanks to ocean currents and **upwelling**, the coastal sea temperature seems to be totally unrelated to the local climate. In Chapter 11 we go a little into why this is so, and discuss the temperature variation of surfing waters around the globe.

In addition to the temperature of the water, the tidal characteristics of a surf spot are something we should know before we go there. For many surf spots, the state of tide means the difference between waves or no waves, and without a set of tide tables, some surfers would be lost and confused. Thankfully, the tides can be predicted fairly accurately, and in Chapter 12, after briefly delving into the mechanics of the tides, we discuss a few ways of predicting their movements.

Knowing all about the water temperature and tidal movements, down to the finest detail, is no good whatsoever if the surf spot you are headed for is going to be flat. The probability of waves is the overriding factor when deciding on a surf destination. In Chapter 13 we look at the **climatology** of the waves – in other words, how wave sizes vary throughout the world at different times of the year.

Chapter 14 provides a brief guide on how to interpret some of the most useful tools available for predicting surf. If you are in any doubt about the meaning of all those symbols and colour contours on the charts shown in previous chapters, turn to Chapter 14 first. It provides a good starting point, and a few tips on what to watch out for when interpreting the charts; but we do not pretend to give a comprehensive guide to wave-forecasting.

The subjects of which this book merely scratches the surface are some of the most conceptually difficult and mathematical. Fluid mechanics, dynamical meteorology and physical oceanography are all involved in finding out how waves are produced and what makes them good or bad for surfing.

Predicting the size of the waves arriving on the coast is one notch up on the difficulty scale from predicting the weather. And predicting how the waves move the sediment, and how this in turn affects the breaking waves, is one notch up again. Add to that the numerous factors that we, the surfers, consider important but which, so far, have been poorly studied by scientists, and it becomes clear that predicting the surf to any great accuracy is going to be extremely difficult.

Knowing every detail about the surf at any given moment – all its characteristics; how it got that way, and how it is changing from one minute to the next – is also extremely complex. So many contributing factors make the waves the way they are – wind, tide, sea-floor topography, swell direction, swell quality and a host of other infinitely variable parameters – that a really comprehensive explanation of why the surf is like it is at any given moment is perhaps impossible. This book does not pretend to give such an explanation, but hopefully it will answer a number of questions about the surf and, in the process, enhance our appreciation of it.

'What's the prediction for Saturday?' asked a third surfer.

'Flat,' answered the first.

'Thank goodness for that,' said the third, 'I've got a wedding on Saturday and I can't go surfing.'

'Who's wedding is it?' asked the first.

'Mine,' said the third.

LARGE-SCALE 2 WEATHER PATTERNS

Introduction

This chapter explores the first stage in our journey. Before a wave is even born, a series of processes go on between the original energy source that is the Sun's radiation; the formation of weather systems, and the generation of waves on the ocean. Through a series of progressively more realistic models of the Earth, we arrive at a global circulation pattern – a framework superimposed upon which are the myriad perturbations and sub-perturbations that make up our day-to-day weather, wind and waves. We also look at the Coriolis force – a fundamental mechanism that makes lows and highs rotate the way they do; a well-known, but little understood phenomenon that makes ocean currents circulate around the globe; coastal water temperatures colder than normal; and tides, instead of just going in and out, rotate around those mystical imaginary points called **amphidromic points**.

Everything on Earth is driven by the Sun's energy.

The global circulation

Everything on Earth is driven directly or indirectly by the Sun's energy. The Sun is like a huge battery, supplying us with a universal source of power through its seemingly inexhaustible process of nuclear fusion.

Not surprisingly, all the energy needed to supply our weather comes from the Sun. The uneven heating of the Earth's surface by the Sun is the first link in a fascinating chain of events leading to tornadoes, thunder storms, blizzards, and those ocean-borne depressions responsible for the waves we ride.

The fundamental cause of all our weather is the fact that the Sun's energy does not heat up the poles and Equator evenly. The Equator is always hotter than the poles. The atmosphere tries its hardest to counteract this disparity by continually attempting to redistribute the heat evenly over the Earth's surface, making the air move around in quasi-regular circulation patterns, which meteorologists call **statistical highs** and **statistical lows**. These average global circulation patterns are a good starting point before we start to explore the intricate, chaotic perturbations that complicate our weather on a day-to-day basis.

The familiar winter pattern of deep depressions over the sea and strong highs over the land, and the summer pattern of large oceanic highs and slack continental lows, is well known to all of us. But how does this come about? What happens to the atmosphere after the Earth's surface has been heated up initially by the Sun's rays?

To investigate this, we will take things step by step – starting with a very simple and highly unrealistic model of the Earth, and then adding various factors to make it progressively more complicated and realistic.

First, we will consider a fictitious Earth, totally covered in water, with no rotation and no seasons. We will see what happens to the atmosphere when it is heated by the Sun. Then we will add the effects of (a) the Earth's rotation about its own axis; (b) the Earth's rotation around the Sun, and (c) the presence of the continents.

A landless, stationary Earth

Imagine an Earth completely covered in water; that did not rotate, and had no seasons. Since the Equator is nearest to the Sun, it would be hotter than the poles. Because the Earth at the Equator is hot, the surface air would expand and become less dense than its surroundings. Therefore, it would rise through the familiar mechanism of **convection**, leaving a large gap, and allowing air to come in and fill that gap. So, on the surface, air would flow continually in towards the Equator from the north and south. Therefore, a giant three-dimensional circulation pattern would be set up in each hemisphere, with air going from the poles to the Equator on the surface; rising at the Equator; travelling polewards, and then sinking at the poles (Figure 2.1).

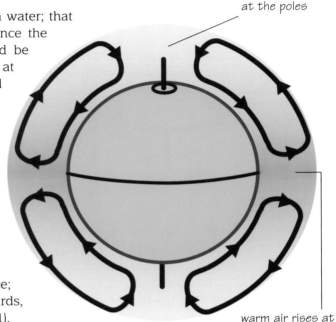

cold air sinks at the poles

warm air rises at the Equator

Figure 2.1 Two-dimensional profile of the large-scale circulation patterns that would occur on a landless, stationary Earth.

A landless, rotating Earth

Now, let us be a bit more realistic and consider the rotation of the Earth about its own axis. This will cause the famous Coriolis force (see page 21) to take effect: all movement in the Northern hemisphere will be diverted to the right, and in the Southern hemisphere, to the left.

Instead of going all the way from the poles to the Equator in a straight line, the surface air will be side-tracked at the first opportunity it gets – it will twist back on itself about a third of the way between the poles and the Equator. The effect of the Coriolis force will have caused the large circulation cells mentioned above to be 'short-circuited' – the result being six spiralling wind-bands, three in each hemisphere.

The air around 60° north and south is continually rising, spiralling upwards. This constant sucking of air away from the surface will reduce the surface pressure locally, which is why the formation of low pressure systems is favourable between 40° and 70° latitude.

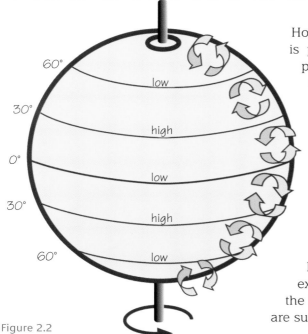

However, the air around 30° north and south is persistently sinking, continually trying to pump more air into the surface layers, and so increasing the surface pressure. This is why quasi-permanent anticyclones, like the Azores High or the North Pacific High, tend to be found at about 30° latitude.

We now have a rotating, water-covered planet with alternate low- and high-pressure belts around its surface, and a weird, three-dimensional spiralling circulation pattern. The general position for the formation of lows and highs has more or less been established, and the existence of the Coriolis force ensures that the lows and highs are circulating the way they are supposed to (Figure 2.2).

Figure 2.2
Spiralling wind patterns that would result on a landless, rotating Earth.

A landless, rotating Earth with seasons

So far, we have a planet covered entirely with water, that is rotating about its own axis once every 24 hours. We have said nothing about its rotation around the Sun every 365 days. If the Earth's axis was sticking straight up and down, then this would not make any difference to our simple model. But in fact, the Earth rotates around the Sun inclined at an angle of twenty-three and a half degrees – an angle referred to by astronomers as the **obliquity of the ecliptic**. Depending on which pole is closer to the Sun, one hemisphere is warmer than the other at any given time of the year (apart from the vernal and autumnal equinoxes – 21 March and 21 September respectively – when both hemispheres briefly receive the same amount of sunshine).

Due to this tilting, the distance between the poles and the Sun varies much more throughout the year than the distance between the Equator and the Sun. This causes wild swings in polar temperatures, but virtually no difference in Equatorial temperatures. In February, for example, the North Pole is much colder than the Equator – perhaps by as much as 70° Celsius. However, in August, the North Pole is quite warm, but the Equator is still the same temperature as it was in February. So, in winter, the difference between polar and Equatorial temperatures is very large, but in summer, this difference is very small (Figure 2.3).

Remembering that air movement is due to uneven heating, we can see that this movement will be massively enhanced during the winter months, due to the extra temperature difference. Those spiralling bands will be working overtime to give us more wind, deeper lows, and bigger surf in winter.

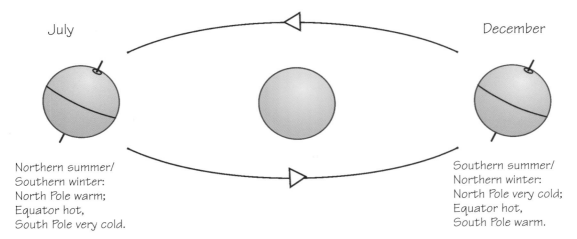

July

December

Northern summer/
Southern winter:
North Pole warm;
Equator hot,
South Pole very cold.

Southern summer/
Northern winter:
North Pole very cold;
Equator hot,
South Pole warm.

Figure 2.3 The
position of the
Earth relative to
the Sun in July
and December,
with its inclination
of twenty-three
and a half degrees.
It can be seen
that, in each hemi-
sphere, there will
be a small Equator-
to-pole tempera-
ture difference in
summer, and a
large Equator-
to-pole tempera-
ture difference in
winter.

A rotating Earth with land, sea and seasons

Until now, we have imagined a planet devoid of all land. Let us get more real, and start thinking about an Earth with continents on it.

The first thing to realize is that land and water respond differently when heated and cooled. Water has a much greater **specific heat capacity** than land – in other words, it takes a lot longer to heat up and cool down. While the temperature of the continents swings many tens of degrees from winter to summer, the oceans keep a more constant average temperature throughout the year, with local variations coming more from coastal upwelling and ocean currents than solar heating differences.

So summer means warm land, not-so-warm sea. And winter means cold land, not-so-cold sea. This adds a new dimension to our six-band system of high and low pressure. The bands are broken up in an east-west direction, and this effect is not the same from winter to summer.

In summer, the pressure over the land is relatively lower than the pressure over the sea. This is because, over the land, the air heats up and therefore rises more. So those quasi-permanent highs that hover around 30° latitude tend to stay over the sea in the summer, leaving low-pressure systems over the continents.

However, in winter, the sea can be warmer than the land. This causes more convection, and therefore lower surface pressures over the sea, with relatively high pressures over the land. This, coupled with the extra levels of energy in the winter, adds to oceanic instability, and feeds all that extra energy into swell-producing depressions.

The picture is now starting to look a little more complex. The last impor-tant thing to add is the fact that there is more land in the Northern hemi-sphere than the Southern, which changes things yet again. The east-west pressure variation, due to the presence of land, is much more noticeable

Figure 2.4
Schematic illustration of 'averaged' isobaric charts for (a) January and (b) July. In the Northern hemisphere, mid-latitude depressions are prominent in winter, but practically disappear in summer, meaning a large summer-winter difference in wave height. In the Southern hemisphere, winter is marked simply by a shifting north and slight intensification of the roaring forties, meaning less summer-winter difference in wave height.

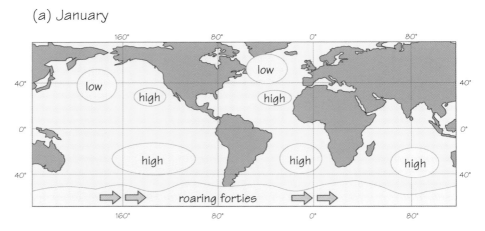

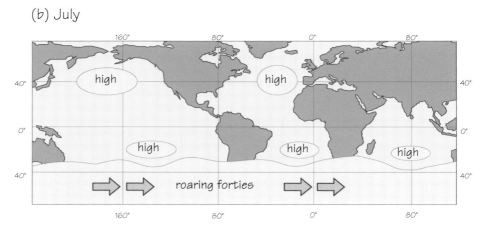

in the Northern hemisphere, simply because it contains more land. The Southern hemisphere suffers a lot less from seasonal variations, and its circulation patterns are much simpler. Typical characteristics of the Southern hemisphere are the **roaring forties** – a band of westerly winds that blow continuously around the globe between 40° and 60° south (associated with that 60° low-pressure band) – and a series of highs just to the north. In summer, the whole pattern shifts south, and the westerlies weaken. In winter, the pattern shifts north; the highs get squashed up, and the westerlies gain strength (Figure 2.4).

The extra simplicity and lack of seasonal variation in the Southern hemisphere means that the waves tend to be more consistent all year round. In the Northern hemisphere, however, you might get huge, unpredictable surf in the winter, and long flat spells in the summer. We will go into this in more detail in Chapter 13.

The Coriolis force

The Coriolis force will be mentioned frequently in this book. A rudimentary appreciation of this strange but vitally important phenomenon is essential to an understanding of such things as lows and highs, oceanic currents, upwelling and tides. In fact, apart from very small-scale processes, almost everything in the world of physical oceanography is affected by it.

The Coriolis force was discovered in the early nineteenth century by the French physicist Gustave Gaspard Coriolis. It is the reason why everything that is moving in the Northern hemisphere turns right and everything that is moving in the Southern hemisphere turns left – the reason why lows and highs, ocean currents and tides circulate the way they do.

We all know that the Earth rotates from west to east, and takes about 24 hours to go around. If you look at the lines of latitude, you can see that the biggest one is the Equator itself, and they get smaller as you go towards the poles. Therefore, if you were on the Equator, you would cover more distance in those 24 hours than you would if you were, say, in London. If you were right on one of the poles, then you would not cover any distance at all – you would just go round and round in the same spot. So, in effect, a point on the Earth's surface is going very fast in an easterly direction at the Equator (about 1,600 km/h), and progressively more slowly towards the poles. The eastward speed of a point at 60° north, for example, is only about 800 km/h.

The best way to begin to understand the Coriolis force is to think in terms of a single 'air parcel', like a kind of balloon. This is embedded in the rest of the atmosphere that is rotating with the Earth's surface. The air parcel is fixed to the surrounding atmosphere, but you can 'unstick' it and move it around. What happens to this air parcel as you try to move it around is the key to the Coriolis force.

Think of the air parcel in London, for example. Before you do anything to it, it is already going from west to east at a certain speed – let's call it 'London' speed. If you now suddenly push it southwards towards Madrid, the air parcel will still be going from west to east at 'London' speed, but the ground beneath it will be going faster ('Madrid' speed). Therefore, the air parcel will lag behind relative to the Earth's surface. Instead of going directly north-south, it will veer off to the right (Figure 2.5). If, on the other hand, the air parcel starts off in Madrid, and you push it northwards towards London, the opposite will happen. The air parcel, having carried with it its initial 'Madrid' west-east velocity, will find itself travelling eastwards faster than the Earth beneath it, and so will tend to swerve to the right.

Now, what happens to the air parcel if you force it in an east-west direction? Say the air parcel is in London again, travelling from west to east with the Earth at 'London' speed. This time you give it a push eastwards towards Moscow. Effectively, you have increased its rotational speed slightly. This increase in speed means an increase in **centrifugal force**, tending to throw

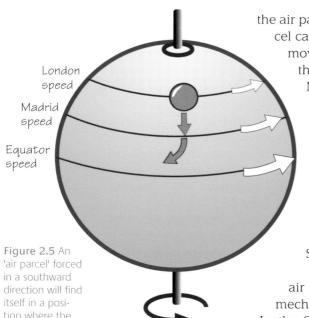

London speed

Madrid speed

Equator speed

Figure 2.5 An 'air parcel' forced in a southward direction will find itself in a position where the eastward velocity of the Earth is increased. It will therefore be diverted to the right.

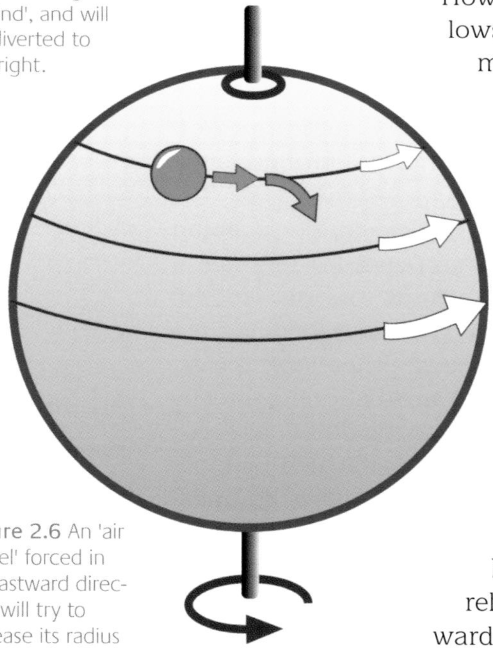

Figure 2.6 An 'air parcel' forced in an eastward direction will try to increase its radius of curvature, and will therefore be diverted towards the Equator.

the air parcel outwards. The easiest thing the air parcel can do to increase its radius of curvature is to move to a position where the Earth is fatter. So the air parcel, instead of going east towards Moscow, ends up swerving south towards the Equator: it turns to the right (Figure 2.6). Conversely, if you give the air parcel a westward push from Moscow towards London, you are effectively decreasing its rotational speed. This will decrease its centrifugal force, and make it want to travel around a smaller radius of curvature, nearer the North Pole. Instead of ending up in London, it veers off towards Scotland: again, it veers to the right.

Therefore, whichever way you push the air parcel, by a combination of two different mechanisms, it will swerve around to the right. In the Southern hemisphere, the whole system is mirror-imaged: by exactly the same principles, the air parcel will always be deflected to the left.

How does the 'air parcel' help us to understand why lows and highs rotate the way they do? First of all, we must look at weather systems as if they were made up of an infinite number of discrete air parcels. The natural tendency for one of these is to try to travel from high to low pressure – 'down the pressure gradient'.

Imagine a centre of low pressure in the Northern hemisphere, surrounded by air at a relatively higher pressure. Immediately there will be a stream of air parcels trying to rush into the low from outside – trying to 'fill up' the low and equalize the pressure. But they will be deflected by the Coriolis force. They will veer off to the right, and end up revolving around the low pressure in an anticlockwise direction (Figure 2.7). The same principle can be applied to a high-pressure cell surrounded by relatively low pressure. The air will try to flow outwards from the high, but will be deflected to the right and end up going around the high in a clockwise direction (Figure 2.8). Again, this can easily be mirrored into the Southern hemisphere, where lows go clockwise and highs go anticlockwise.

It should now be clear that the Coriolis force is absolutely fundamental in the world of meteorology and oceanography. It not only makes highs and lows rotate the way they do, but also, through a series of cascading effects, it is the reason why ocean currents bring cold, polar water to west-facing coasts; why there is high biological productivity along the coasts of Peru and Chile, and why tides behave in such strange ways.

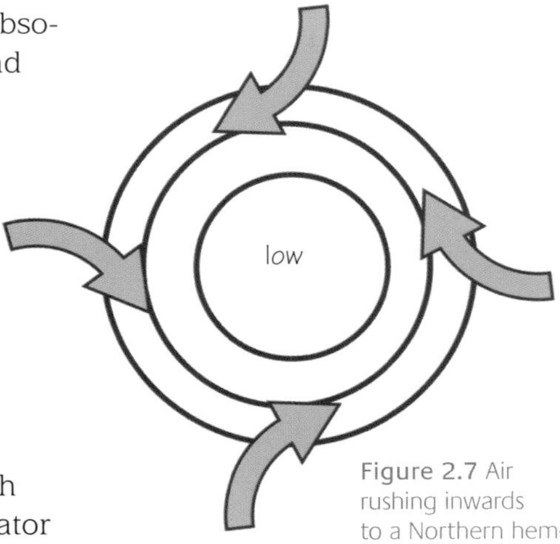

Summary

The fundamental reason why we have storms on Earth is the fact that the Sun heats up the poles and the Equator unevenly. Heat has to be transferred between the two.

In this chapter we have described a general global circulation pattern, through a series of hypothetical models of the Earth. We have seen that, if the Earth were landless and stationary, the air would simply rise at the Equator, travel towards the poles and then sink. In fact, the Earth rotates; these large circulation patterns are 'cut-off' by the Coriolis force, and the result is a six-band system of alternate low- and high-pressure zones. The Earth's seasons greatly enhance the air movement in the winter hemisphere, due to the increased Equator-to-pole temperature difference. However, we must remember that the Earth has continents; it is not covered in water. This causes the six-band system to be broken up from east to west. In summer, there is a high over the sea, and a low over the land; and vice-versa in winter. This, coupled with the fact that the winter hemisphere contains extra energy anyway, means that the favourable time for deep lows over the sea is winter.

This chapter has also looked at that vitally important phenomenon, the Coriolis force, which dictates why large-scale motions on the Earth's surface, like highs and lows, rotate the way they do. The deflection to the right in the Northern hemisphere can be explained first by the fact that a north-south travelling object will arrive at a position where its west-east velocity is less than that of the ground beneath it. Second, a west-east travelling object will already have a greater velocity than the ground beneath it and, due to the fact that its centrifugal force will exceed gravity, it will increase its radius of rotation, and end up nearer the Equator. Air flowing from a high pressure to a low pressure will therefore turn to the right, so that, in the Northern hemisphere, air around a high will rotate clockwise, and air around a low, anticlockwise.

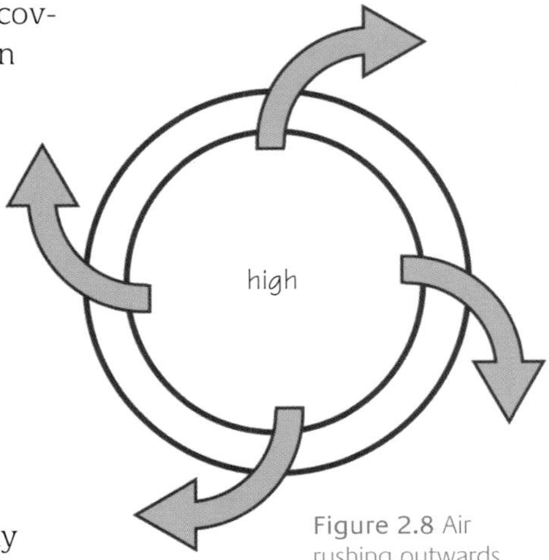

Figure 2.7 Air rushing inwards to a Northern hemisphere low pressure is diverted to the right, resulting in the system rotating anticlockwise.

Figure 2.8 Air rushing outwards from a Northern hemisphere high pressure is diverted to the right, resulting in the system rotating clockwise.

3 THE FORMATION OF A DEPRESSION

Introduction

Without doubt, the most important phenomenon for producing the waves we ride is the **low pressure**, also called the **mid-latitude depression** or **extra-tropical cyclone**. The low pressure is really just a cell of air whose pressure is lower than its surroundings. However, thanks to the ever-present Coriolis force, it also features a swirling pattern of fast-moving surface air, which generates waves on the sea by frictional drag on the water surface. The deeper the depression, the faster this air moves. The faster it moves, the more it drags on the water and the bigger the waves will be.

But how does this phenomenon come about? What are the factors that combine to allow one of these systems to come into being in the first place? In this chapter we will look at the chain of events leading to the formation of a large, swell-producing low pressure.

Beginnings

Satellite photo of a low in the Southern hemisphere. South-east Australia is visible.

Some of the first people to study the formation of a low pressure were a group of Scandinavian meteorologists from Bergen in Norway, headed by Vilhelm Bjerknes. They came up with the concept of the **polar front** – a band on the Earth's surface where cold air from the poles meets warm air from the Equator, and where the spawning of the mid-latitude depression takes place. The position of the polar front coincides with that general band of relatively low-surface pressure in the large-scale circulation patterns described in Chapter 2. It is the boundary between two circulation cells, where the surface air is flowing in opposite directions (Figure 3.1).

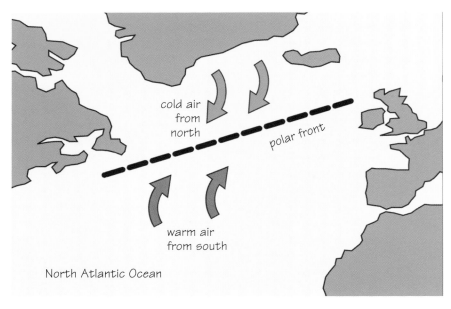

cold air
from
north

polar front

warm air
from south

North Atlantic Ocean

Figure 3.1 Typical position of polar front in the North Atlantic, where air masses meet from the north and south.

In the Northern hemisphere, for example, the polar front is where cold air coming from the north meets warm air coming from the south. These 'blocks' of air with different physical characteristics – in this case, temperature – are called **air masses**. Warm air is less dense than cold air; therefore, in a polar front, the warm air tends to slide over the top of the cold air, and this generally happens before any kind of low even begins to develop.

Through a particular combination of circumstances, a disturbance may appear at some point along the front. For example, the north-south air temperature difference may be particularly intense at this point, or there might be some influence from an 'external' factor like the sea surface temperature. Such a disturbance is known to meteorologists as **baroclinic instability**. If the perturbation is strong enough, and all the factors are right, the front will develop a 'wave' on it, which will grow and intensify. The process of warm air sliding over the top of the cold air will be particularly intense in the area of the perturbation, so this will lead

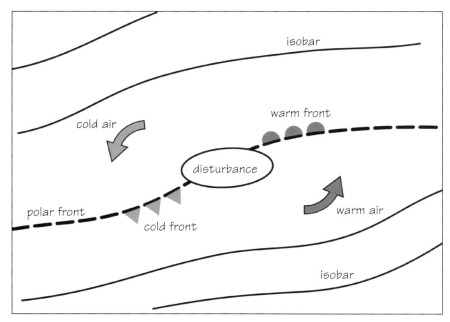

isobar

warm front

cold air

disturbance

polar front

cold front

warm air

isobar

Figure 3.2 A small disturbance on the polar front develops into the beginnings of a low-pressure system. Warm and cold fronts develop on the polar front, ahead of the warm and cold air masses. If conditions are favourable, the system will begin to spin anticlockwise.

to a localized drop in surface pressure. Air will then start to rush in from outside, deflected into a vortex by the ever-present Coriolis force (Figure 3.2).

The fully developed low

As the warm air slides over the cold air, and the baroclinic perturbation begins to grow in intensity, the whole thing turns into a huge vortex: the air being furiously sucked in and deflected by the Coriolis force, resulting in an anticlockwise rotation (in the Northern hemisphere – clockwise in the south). This section of the polar front can now be seen to split into a system of individual fronts – the **warm front**, behind which is warm air, and the **cold front**, behind which is cold air. Between these two fronts is the **warm sector** – an area where the surface winds are strong and blowing in the same direction for some distance – the best conditions for rapid wave growth (Figure 3.3).

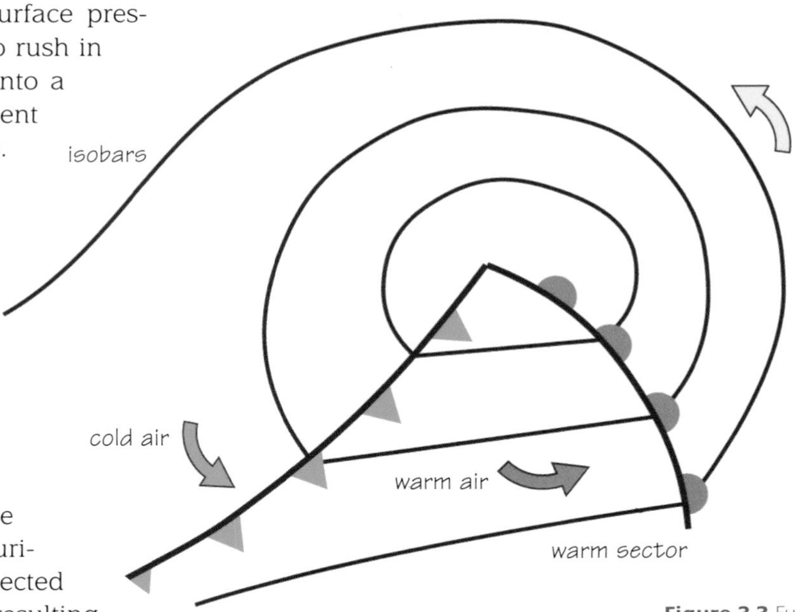

isobars

cold air

warm air

warm sector

Figure 3.3 Fully developed low-pressure system, with warm sector between the warm and cold fronts.

This is the fully developed low-pressure system. If all the factors are right, and the surface pressure manages to reach a very low value, then the wind will be strong and the waves big. On the chart, a deep low can be recognized straight away by a thick mass of closely packed **isobars**. The closer together they are, the lower the surface pressure inside the depression; the quicker the air tries to rush in from outside, and therefore, the stronger the surface wind. Because the air trying to rush in is deflected almost 90° by the Coriolis force, the direction of the wind is more or less along the isobars.

As the depression propagates from west to east, the warm and cold fronts eventually catch up with each other, forming an **occluded front**. This is when the low starts to weaken and, ultimately, maybe after spawning a few peripheral systems, lose its identity altogether.

But it is before this occlusion starts to take place, when the low is in its fully mature state, with warm and cold fronts fully separated and winds blowing strong and straight in the warm sector, that it is of most use to us. The longer it stays like this, and the more it hangs around over the same stretch of ocean, the bigger the waves will be. A very deep low that stalls in its path is rare, but if this happens, it can pump out huge, clean swells for days. Normally, when lows are deep, the atmosphere contains a lot of energy and a low moves fast, making way for its successor to appear hard on its heals – a situation not altogether bad, but the whole ocean may end up with a criss-cross of different swell directions.

Explosive cyclogenesis

Some lows deepen to phenomenally low values. The deepest one ever recorded, at 914mb, was the Braer Storm of 10 January 1993, which was well documented because, sadly, a disaster was associated with it. This was the final destruction of the *Braer*, a large oil tanker stranded on the rocks in the Shetland Islands by a previous storm. The phenomenon of a low developing extremely quickly – more specifically, one that deepens more than 24mb in 24 hours – is referred to as **explosive cyclogenesis**.

The millennium swell

Figures 3.4 to 3.6 show a sequence of Atlantic isobaric charts, with a clear example of the development of a low from the slight disturbance on a frontal system, to the fully mature and occluding low pressure, in the space of two days. This is an extreme example – the system deepened very quickly indeed, but it also moved rapidly in a north-easterly direction. The constant broad band of westerlies on its southern flank ensured huge surf for most of Europe during this time. Chapter 14 explains in more detail how to interpret charts like these.

Upper air influence

In the charts shown you will have noticed that the low moved extremely quickly north-east and deepened rapidly. There is a link between these two phenomena. The movement of a low pressure has a lot to do with the flow of air about 5,000 metres up – the **jet stream**. Members of the **Bergen school** discovered that the behaviour of the upper air affects the pressure on the surface through three-dimensional convergence (squashing) and divergence (stretching), combined with the effects of the Coriolis force.

The airflows 5,000m up are much stronger than those on the surface, and a little less complicated due to the lack of influence from the land. The jet stream is a constant westerly wind that blows around the globe at mid-latitudes, 5,000m up. But it has meanders in it, which make the flow waver north and south. The position and size of these meanders vary constantly.

One of the discoveries made by the Bergen school was that the path taken by a surface low closely follows the track of the jet stream. The amount of energy pumped into the low by the upper air stream depends on how strong

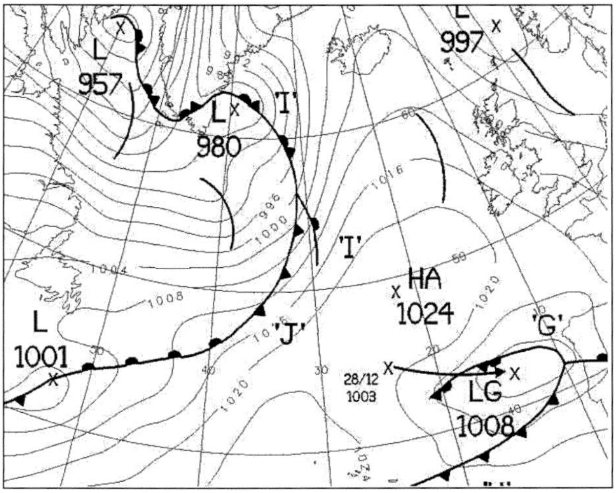

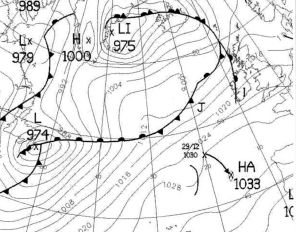

Figure 3.4 00h, 29 December 1999. The birth of a depression. In the bottom left-hand corner of the picture a front has formed, to the south of which lies warm air and to the north, cold air. There is already a centre of relatively low pressure (1001mb), around which air is beginning to circulate.

Figure 3.5 00h, 30 December 1999. Twenty-four hours later, the frontal system has split into its warm- and cold-front counterparts. The warm sector, with its straight isobars, can be seen between the two. The low has deepened considerably (to 974mb) and moved north slightly.

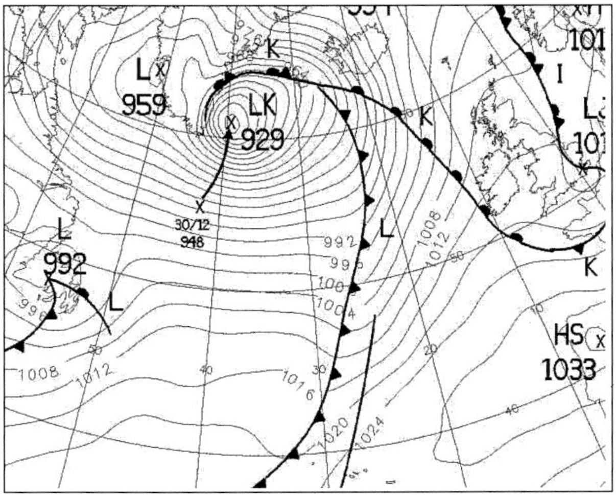

Figure 3.6 00h, 31 December 1999. Explosive cyclogenesis. The low has moved rapidly north-west, and deepened an incredible 45mb in the last 24 hours (to 929mb). It is now just past its fully mature state, and has already started to occlude. The tightly packed isobars on the southern flank of the low mean storm-force westerlies and huge surf.

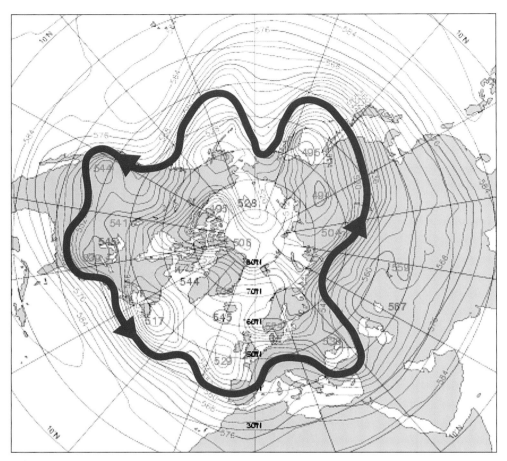

Figure 3.7 Upper air flow chart of the Northern hemisphere. The thick meandering line represents the approximate position of the jet stream 'core', or area of maximum wind speed.

the jet is, and also on its orientation. The deepest lows tend to develop in places where the jet stream is very strong, and on the poleward arm of a meander (Figure 3.7). For example, in the Atlantic, the formation of deep lows is most favourable when the jet is flowing diagonally in a north-easterly direction across the ocean. Lows form somewhere off the eastern seaboard of the USA, and then deepen rapidly while tracking towards Scotland and Ireland. However, if the jet is particularly weak; split into two paths, or orientated from north-west to south-east across the ocean, then the formation of surface lows is not so likely. In the Pacific, lows tend to form between Japan and the Aleutian Island chain. In Figure 3.7, it can be seen that the jet stream over the Pacific happens to contain two distinct north-easterly flowing arms, where the formation of low pressures is more likely.

Summary

In this chapter we have looked at how a mid-latitude depression – by far the most important source of waves for surfing – develops from a small atmospheric disturbance into a full-blown, raging storm. The system is spawned on the polar front – a band around the Earth where warm and cold air masses meet. If conditions are right, a localized perturbation on this front grows into a vortex through a chaotic chain of events. The system then develops into a mature depression – a rotating mass of surface air that propagates from west to east across the surface of the ocean.

The development of the mid-latitude depression, from the initial wave disturbance, through to what can be a devastating storm, is an excellent example of a small change in initial circumstances leading to a massive difference in final results through a chaotic series of knock-on, or cascading, effects. Eventually, the system gives up and dies, but not before it imparts much of its energy to the sea's surface, generating a mixed-up and seemingly random series of waves of all shapes and sizes, which will eventually organize themselves into clean lines of swell propagating towards our coast.

Box 1 The geostrophic equation

$$\text{windspeed} = \frac{1}{\text{Coriolis}} \times \text{Pressure gradient}$$

This equation shows that, since the Coriolis force increases with distance from the Equator, for the same pressure gradient, windspeeds will be greater near the Equator. Therefore, the same low will have stronger winds at lower latitudes.

To calculate the windspeed from an isobaric chart, we can use the following form of the geostrophic equation:

$$\text{windspeed} = \frac{690}{\sin(\text{latitude})} \times \frac{\text{pressure between isobars}}{\text{distance between isobars}}$$

For the windspeed to come out in metres per second, the latitude must be in degrees, the pressure difference between each isobar in hPa and the distance between isobars in kilometres. Note that we have assumed the air density to be $1\,\text{kg/m}^3$.

For more information see:
Ahrens, C.D., 1999, *Meteorology Today: An Introduction to Weather, Climate, and the Environment* (6th edn), Brooks/Cole Publishing.

4 THE GROWTH OF WAVES ON THE OCEAN

Introduction

In Chapter 3, we came to the conclusion that a fully developed, mid-latitude depression, with its swirling vortex of surface winds blowing over the ocean, is the most common source of swell. The waves themselves are generated by nothing more than the action of the wind blowing over the surface of the sea – the air dragging on the water surface. Energy is transferred from one medium to the other, the motion of the air converting a smooth and static sea surface into a ruffled-up and dynamic wave-field.

In this chapter we will look at how this energy is transferred from the air to the water – in other words, how waves are produced from the action of the wind blowing over the surface of the ocean. We will also go into a little of the history behind modern techniques for predicting waves – the predecessors to the massive 3G **WAM** model – without which the lives of all budding surf-forecasters and internet geeks alike would totally fall apart.

The Miles-Phillips theory

Exactly how the air transmits its energy to the water to make waves is still not completely understood. But since the work of legendary oceanographers J.W. Miles and O.M. Phillips in 1957, things have become much clearer. It is generally agreed that there are probably two mechanisms involved. The first starts by producing small waves from a completely flat sea, and then, once these small waves have established themselves, the second mechanism can take over. The second mechanism goes to work on the small waves to make them into bigger ones, and then makes the bigger ones into even bigger ones, until some kind of limit is reached whereby they cannot grow any more.

Capillary waves

At the initial stage of wave generation are the little bumps called **capillary waves**. They are produced from a completely flat sea. The key to how this happens is the fact that the wind does not really blow completely horizontally all the time. It naturally contains little random disturbances that make it go up and down as well. Sometimes these little disturbances are enough to make the air start pushing down on the sea, making tiny up-and-down motions on the surface of the water itself. This is the vital beginning, the catalyst needed to trigger further reactions between wind and water, facilitating the flow of energy and allowing bigger waves to grow.

These mini-vortices tend to drift about like little swirly tornadoes turned on their side, and, every now and again, one of them happens to follow the wave it just produced. This adds further energy to the wave, and makes it grow even more.

A good way to understand this is to use the following analogy. Imagine you have a special hairdryer, which can switch itself on and off about once a second. You point it down, on to the surface of the water. The air pushing down on to the water will cause a depression in the water surface directly below the jet of the hairdryer, and a lump all around where the water has been displaced to one side. When the hairdryer switches itself off, the depression will 'spring back' upwards, and the lump will spring back down. A wave motion will result, which will propagate outwards, away from the hairdryer. Now, if you follow one of the waves with the hairdryer, the wave might grow bigger and bigger as you pump more and more energy into it (Figure 4.1).

The growth rate of waves produced in this way happens to be linear. They grow bigger and bigger at a steady rate with time. Once enough of these little mini-

Figure 4.1
Analogy of linear wave growth mechanism. Mini-vortices (similar to a hairdryer, for example) follow a depression in the sea surface, making the depression progressively bigger, and eventually leading to wave motion.

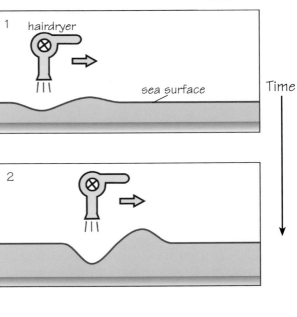

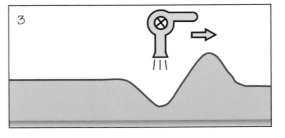

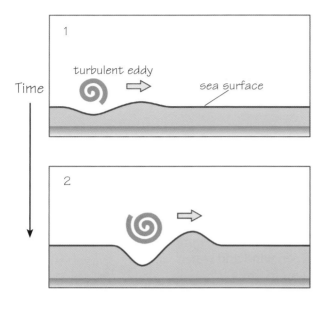

Time

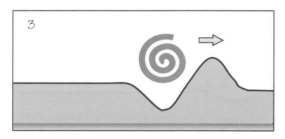

Figure 4.2
Exponential wave growth. The turbulent eddies increase in size at the same time as the wave motion, causing the waves to grow at an ever-increasing rate.

vortices have chased after the waves they have produced, and formed centimetre-high ripples, the sea surface will have a certain 'roughness' to it, and be ready for the second mechanism to take over.

Exponential growth

The second mechanism is self-perpetuating. As soon as some of these little capillary waves start to exist, the surface becomes ruffled. This modifies the air over the surface, producing larger vortices called **turbulent eddies**. These new vortices, instead of being random, are linked to the waves themselves and automatically follow them along. The waves will grow because the vortices increase the pressure over the troughs of the waves and decrease the pressure over the crests. As long as the wind blows over any existing waves, the size of the turbulent eddies will increase, which makes the waves grow even more, which makes the eddies even bigger – and so on.

Put more simply, as the surface gets waves on it, and therefore gets rougher, the wind is able to grip the surface more, so the waves get bigger, which gives the wind even better grip, and makes the waves grow even more quickly. It is an accelerated growth mechanism, exponential with time, not linear. The bigger the waves, the quicker they grow (Figure 4.2).

The waves are no longer capillary waves; they are now called **gravity waves** – so-called because their **restoring force** is gravity. The restoring force is that which restores the sea surface to its original position after it has been lifted by the air motion. With capillary waves, the restoring force is not gravity, but surface tension – the 'skin' that is always present on any water surface.

Limiting factors

These bigger waves now have the ability to get very big indeed. But they do not keep on growing as long as the wind blows. If this were the case, in the roaring forties, which practically continue to blow all the year round, waves would be infinitely large. Obviously, this is not so. When the waves get to a certain height, a balance is reached between the generating force (the wind) and the restoring force (gravity). In other words, the wind cannot push them up any higher, because gravity keeps pulling them down. If the wind is stronger, then they will get higher. Another factor is **whitecapping**, or 'white horses'. This takes away a lot of the energy. In a force twelve, for example, much of the wind power is used up in blowing around giant chunks of foam.

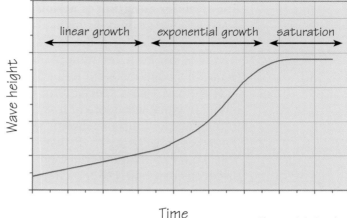

Figure 4.3 Graph of wave height plotted against time. At first the waves grow at a linear rate, then exponential growth takes over. Lastly, due to various limiting factors, the waves reach saturation point and do not grow any more.

Imagine a wind starting to blow over a stretch of ocean. If it keeps blowing, eventually a time will be reached when the waves will not grow any more. But what if the wind stops before they reach their maximum size? This is called a **duration-limited** sea. Alternatively, if the distance over which the wind is blowing – the fetch – is not very big, then the waves will never reach their maximum height either. This might happen in a restricted area like the North Sea, for example. This is called **fetch limited**. Lastly, if both the fetch and the duration are really long – infinite, like the roaring forties, for example – then we have a **fully arisen sea** (**FAS**). This is when the waves are allowed to reach their natural maximum height for the wind strength (Figure 4.3). We shall return to this in Chapter 9.

Empirical models, parametric models, and real physics

The limiting factors mentioned above were all very elegantly incorporated into one of the first mathematical models for wave forecasting. This was invented in the 1940s by pioneering oceanographer Walter Munk and his colleagues, to help with landing operations in the Second World War. This was way before the work of Miles and Phillips, when little was known about how waves were generated by the wind. But it didn't matter. It was an **empirical model** – a way of relating wind strength, fetch and duration, directly to wave height. It was as if there were a magic box, into which you put the values of wind strength, fetch and duration; if you then turned a handle, out would come the answer – the wave height.

What went on inside the magic box was also a mystery, and research was already underway to find out. After the breakthrough of Miles and Phillips in 1957, a special equation was developed, which forms the basis of all wave prediction models used today. The equation – sometimes called the **radia-**

tive transfer equation or **action balance equation** – was used to work out how big waves would be, taking into account not only the energy input from the wind according to the Miles-Phillips mechanism, but also how much energy is lost from, for example, whitecapping and friction. The first models to use this equation were termed first generation (1G) wave models.

There was also a special term in the equation, which nobody could fathom, and which was more or less ignored in these first models. It referred not to the energy transfer between wind and waves, but to the energy transfer that takes place between the waves themselves as the sea continues to grow. This was a phenomenon known as **non-linear transfer**. Its existence was discovered around 1963, by a genius called Klaus Hasselmann, of the Max Planck Institute of Meteorology in Germany.

In 1973, Hasselmann and his colleagues performed a huge experiment called the **Joint North Sea Wave Experiment** (**JONSWAP**), to try to find out more about what went on during the wave generation process. One result of this experiment was the surprising discovery that the distribution of wave sizes and shapes was more or less the same in any growing windsea, although obviously some seas got bigger than others. It was found that this discovery could be used to compensate for the inability to compute the difficult part of the radiative transfer equation. This was a technique known as **parameterization**. Out of this came the second generation (2G) wave model.

Do not forget that, throughout this time, computers were becoming increasingly powerful and that, since calculating the height of waves on the sea surface requires the biggest computers in the world, scientists were always waiting for 'the technology to become available' before they could put their latest theories to the test. By 1985, a special group of scientists had been commissioned to design and put into action the definitive wave prediction model – one that could be used practically, and could be run on fairly 'normal' supercomputers. Three years later, they came up with the WAM model – the third generation (3G) model, variations of which are still used extensively today for wave forecasts throughout the world.

Figure 4.4 shows an example of a contour map of wave heights and directions obtained from the US Army Corps of Engineers Fleet Numerical and Oceanographic Centre (FNMOC) via the internet. The model used to generate the forecast was the Wavewatch III, a third-generation wave-prediction model. We will look more closely at wave modelling in Chapter 14.

Summary

In this chapter we have seen how the action of the wind blowing over the surface of the water imparts its energy upon the ocean and generates waves. This is the only thing that produces the waves that eventually we will ride.

Exactly how the energy is transferred from the air to the water is still not completely understood. However, since the late 1950s we have accepted

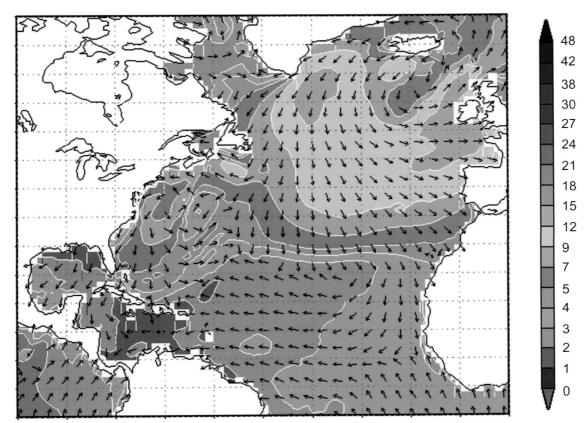

Figure 4.4 Example of wave height contour chart from the Wavewatch III model, obtained on the internet from FNMOC. Colours indicate wave height in feet, and arrows indicate swell direction.

that it is by a combination of two mechanisms. First, capillary waves are formed from a flat sea by mini-vortices in the wind flow. Then, through a feedback mechanism between the wind flow and the waves themselves, the growth soon becomes exponential, the capillary waves transforming into full-blown gravity waves. But waves cannot grow forever: they are limited by the strength of the wind; the distance over which it blows, and how long it blows.

We also looked briefly at the history of wave forecasting. The first empirical models for forecasting waves were invented around the time of the Second World War, in the 1940s. Then, after the pioneering work of Miles and Phillips in the 1950s, the mathematical basis behind modern wave prediction models was established. A form of the third generation model, published by the WAM group in 1988, is still being used today.

The waves described here are strictly those within the generating area itself – the storm centre. They are the waves that exist directly beneath all that wind: a mixed up, seething confusion of random wave motions – a windsea. These waves are of all shapes, sizes and directions, existing at the same time, exactly as we get on the coast when the wind howls onshore. (See note on page 38.) But after being generated in the area of the low pressure itself, the waves propagate out into other waters, no longer having energy input to them by the wind. This is free-travelling swell, and its peculiarities will be described in Chapter 5.

Box 2 The action balance equation

The *action balance equation* or *radiative transfer equation* is the fundamental equation used in modern wave prediction models. It equates the growth and evolution of the waves with the input of energy from the wind; the dissipation due to things like whitecapping and opposing winds, and the transfer of energy between the waves themselves:

For more information see:
Komen, G.K., Cavaleri, L., Donelan, M., Hassleman, K., Hassleman, S. and Janssen, P., 1994, *Dynamics and Modelling of Ocean Waves*, Cambridge University Press.

Note that some surfers who live in limited fetch areas do not have the luxury of groundswells that have propagated thousands of kilometres away. They have to surf virtually in the storm-centre itself. This is covered in detail in Chapter 9, 'Surfing in the Storm'.

5 PROPAGATION OF FREE-TRAVELLING SWELL

Introduction

So far, we have looked at how waves are produced by wind blowing across the surface of the ocean: wind that is the result of air trying to rush from a region of high pressure into a region of low pressure. The area over which the wind is blowing – the wave-generating area – is more or less the location of the depression itself, and is referred to as the storm centre. Here, the waves are constantly supplied with energy by the moving air, and the sea contains a wide range of waves of all different sizes, lengths, shapes and directions. This is a windsea – basically, a blown-out stormy sea.

When the waves start leaving the generating area, they no longer remain under the influence of the overlying wind, and propagate away as free-travelling swell (Figure 5.1). Very little energy is actually lost when swell travels on its own, and although the waves get progressively smaller, as described below, they can be detected literally thousands of kilometres away from the storm centre. For example, in 1957, Walter Munk managed to detect waves

wind input to waves here

storm centre

free-travelling swell (no wind input)

Figure 5.1 Inside the storm centre the wind transfers its energy to the waves. Outside the storm centre, the waves are free-travelling, no longer under the influence of the wind.

at Guadalupe Island off the west coast of Mexico that were generated over 15,000 kilometres away in the Indian Ocean. Also, a group of scientists led by D. Snodgrass in 1966, performed an amazing experiment in which they used a series of measuring stations to track swells generated near Antarctica, all the way across the Pacific Ocean to Yakutat, Alaska – a distance of over 10,000 kilometres.

If the storm centre is some distance away, then the waves we ride on the beach are quite different from those originally generated. Several things happen to the swell as it propagates away from the storm centre (Figure 5.2), and later in this chapter we will investigate the three most important ones:

1 *Circumferential dispersion:* the waves get smaller as they spread out over a larger and larger area.
2 *Radial dispersion:* some of the waves propagate faster than others, the faster and longer ones arriving on the coast first.
3 *Grouping:* throughout the propagation path, the waves sort themselves into sets – a vital phenomenon familiar to all surfers, but extremely difficult to analyse.

The strange characteristics of swell

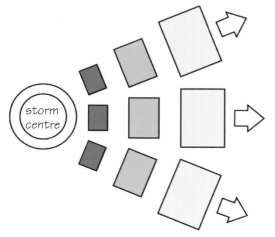

storm centre

Figure 5.2 Schematic illustration of the swell propagating away from the storm centre, depicted as a series of 'packets'. The further from the storm centre each packet gets, the more it expands in both radial and circumferential directions.

When we think of waves travelling over the surface of the ocean, we must remember that, in deep water anyway, waves do not 'carry' the water from one place to the other. They are not like ocean currents, where water from the Antarctic can end up in the Atlantic. Waves are simply carriers of energy. This is easily demonstrated if you get a piece of carpet and flick it up and down so that a wave propagates from one end to the other. The wave you generate does not 'carry carpet' from one end to the other; it simply transfers the up and down motion of the carpet from one end to the other. The sea surface is just like a carpet, but it is joined to the water beneath it.

When waves propagate over the sea, they transfer the up and down motion of the surface from one place to another. In fact, the motion is not just up and down, it is circular, or orbital. If you looked at any floating object from the side, you would see it do a complete circuit every time a wave passed through, ending up in more or less the same spot (Figure 5.3). This orbital motion is transferred to the water below the surface. Because of friction and energy loss between vertical levels in the sea, the orbits of the particles within the water get smaller the deeper they go, eventually diminishing to nothing at some depth (equal to approximately half the wavelength) below the surface.

In this chapter we are concerned only with really deep water, where the height of the wave is insignificant compared with the depth of the water. But it is worth mentioning that, in shallower water, the orbital motions of the water particles start to get a little more complex. First, in shallower water the height of the wave becomes significant compared with the depth of the water. Because wave speed depends on depth, the top of the wave goes faster than the bottom. Eventually, the wave topples over and breaks, but before this happens, the orbital motions change slightly, so that the top part of the orbit is faster than the bottom part. The result is a slight forward displacement of water, known as **Stokes drift** (Figure 5.4). Second, the water is not deep enough for the orbital motions to diminish to nothing – they hit the bottom before this happens. This squashes the circular motions into ellipses. Then right on the bottom, all that is left is a backwards and forwards motion of the water next to the sea-bed. This is very important for the movement of sediment (see Chapter 8).

In deep water, when a wave propagates over the surface, all it carries with it is a message. The message, passed along the surface, simply tells the surface to go round in an orbital motion. The fact that waves are messages is the fundamental physical principle behind sound, light, radio, and almost every kind of communication method known. When Walter Munk detected those waves from 15,000 kilometres away, what he detected was a certain pattern in the orbital motion that he identified as coming from the Indian Ocean.

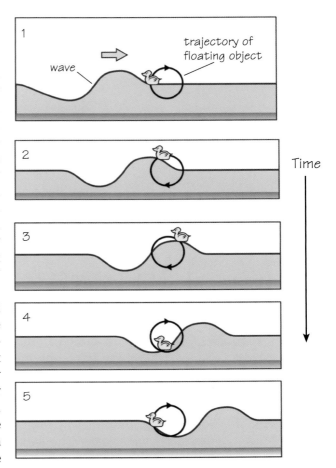

Figure 5.3 In deep water, a floating object traces an orbital path as a wave passes through. The object ends up in more or less the same place.

deep water

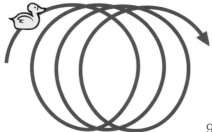

shallow water

Figure 5.4 In deep water the floating object has no net movement, but in shallower water there is a slight net forward motion called Stokes drift.

However, the waves that Professor Munk saw in Guadalupe were not the same waves that started off in the Indian Ocean. Strange as it may seem, a single wave cannot propagate very far on its own, but a group can. Later on in this chapter, we will look at how the waves organize themselves into groups or sets, but here let's just imagine we already have a group with six waves in it. This group moves across the ocean at a speed called **group speed**, and it is here in the group where the energy (and the message) is carried. But each individual wave in the group travels at twice the speed of the group – a speed known as **phase speed**. As the group crawls along the surface of the ocean, all the waves in it are constantly moving from the back of the group to the front. This is a bit like taking a moving staircase, laying it flat on the ground while it is still going, and then pulling the whole thing along. Each stair would appear at the back of the staircase, travel to the front, and then disappear. This is what happens with waves in a group. Waves are generated at the back of the group, move through the group to the front, and then disappear (Figure 5.5).

Circumferential dispersion

As the swell travels away from the storm centre, it spreads out over a progressively wider area. The width it spreads out depends on the distance it has travelled – the further it goes the more it spreads out. The swell becomes less concentrated as it travels, so the waves get smaller. This is **circumferential dispersion**.

If the storm centre is very concentrated or the fetch is very narrow, the waves could be considered to be generated from a single point source. In this theoretical case, for every doubling of the distance from the storm centre, the energy of the waves reduces by half. Owing to the relationship between wave height and energy, this means that the wave height would reduce by about 30 per cent. In practice, storm centres are not singular point sources and fetches are often quite wide, which has the effect of making the reduction in wave height smaller, often around 15 to 20 per cent.

As a theoretical example from the north Atlantic, let us think of a low pressure sitting just south of Iceland, pumping out swell towards western Europe. For simplicity, we will assume this low pressure is like a point source. As the swell leaves the storm, it will start spreading out. By the time it reaches a spot, say, 1,000km away – somewhere in Ireland perhaps – the waves arriving on the coast might be 5m high. Some of the waves end their life here, giving up their energy as they break on the reefs and beaches of

Ireland. However, a great deal of this swell continues on its journey southward. When it reaches a point, say, 2,000km from the storm centre – the north coast of Spain, for example – the height of the waves will be 30 per cent less than they were in Ireland (about 3.5m). When the waves have travelled another 2,000km – to the Canary Islands, for example – they will have lost another 30 per cent and be down to about 2.3m. (Figures 5.6 and 5.7.) Note that, if the fetch is quite wide, the waves in Spain and the Canary Islands could be somewhat bigger. In the Indian and Pacific oceans, fetches tend to be wider than in the Atlantic, therefore the reduction in waves' height with distance is even less.

Radial dispersion

Water waves are complicated, due partly to the fact that they do not all travel at the same speed. Their speed is governed by their **wavelength**, or how far apart one wave is from the next. The longer the wavelength, the faster they go. A good analogy for this might be a group of marathon runners, some of them having longer legs than others, and therefore being able to run faster. Having all started off together, the longer-legged ones will eventually pull out in front, leaving the shorter-legged ones behind. This is **radial dispersion**.

When a swell is first generated in the storm centre, all sorts of wavelengths are produced at the same time, so very close to the storm centre we see a mixture of many waves of different sizes, shapes and wavelengths. As the swell begins to propagate away from the storm centre, the various different wavelengths begin to sort themselves out, the longer, faster ones racing out in front, and the shorter, slower ones lagging behind. By the time the swell is a long way from the storm centre, the longer waves have made their way right out in front, and the shorter ones have been left way behind. The swell has 'stretched out', but this time in a radial direction, not a circumferential one.

Therefore, if you were sitting very close to the low pressure (Ireland in our example), you would see the whole swell arriving in a very short time. Although the long waves would arrive first, the short ones would not be very

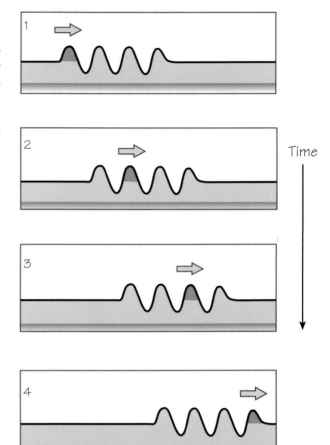

Time

Figure 5.5 Each individual wave in a group propagates at twice the speed of the group, resulting in a constant flow of waves from the back of the group to the front. Individual waves do not last very long – they are born at the back of the group and die at the front.

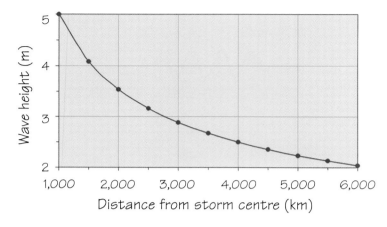

Figure 5.6 A graph to show how circumferential dispersion causes wave height to diminish with distance from the storm centre.

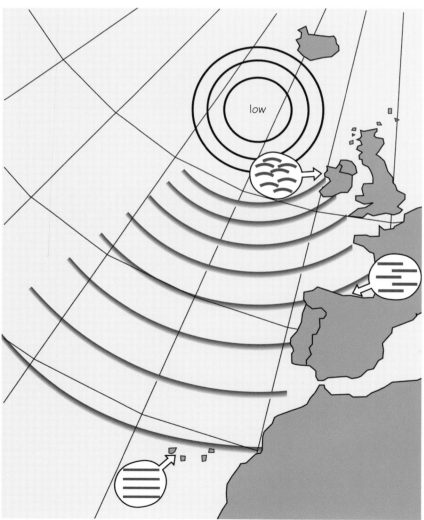

Figure 5.7 A swell generated just south of Iceland will gradually lose height, but will also clean up, as it progressively reaches the shores of Ireland, Spain, and eventually the Canary Islands.

far behind. The effects of radial dispersion would have had little chance to act on the swell, and so the sea might still appear rather confused.

However, if you were, say, 2,000km away (Spain, for example), clearly the long waves, being way out in front, would arrive before all the others. Then all the other waves would turn up, the shortest ones coming in last. The swell would be cleaner and more lined up, with a smaller number of different wavelengths arriving at the same time. Then, even further away (say 4,000km at the Canary Islands), the swell would be even more 'stretched out' radially. Here, the different wavelengths would be completely separated from each other (Figure 5.8).

The first waves to arrive would have more punch to them, because they would be faster, but the biggest ones normally arrive a short time later. The arrival of a new swell is always very interesting to watch. The trick is to be there right at the start; to recognize the lines coming in fast and well-separated, and be ready for the big swell to hit within a couple of hours. Lines will start appearing, straight and regular, barely detectable at first, but with a massive separation between one wave and the next. All you have to do now is wait for it to pick up.

After the swell has peaked in size, the waves do not have the same punch as they did before it peaked. Right at the end of the swell, when the very last waves are coming in, they are normally relatively weak. Sometimes the swell peters out quickly, not lingering too long, especially if something has happened along the propagation path to remove the shortest waves. Often these waves are eliminated, for example by opposing winds in the propagation path, or they might be affected by whitecapping (white-horses). Short waves tend to be affected much more by both these phenomena, because they are steeper and 'stick up' more from the sea surface. This is especially noticeable near the storm centre.

Grouping

Somewhere in the propagation path, between the storm centre and the coast, the waves organize themselves into sets – groups of two, three or maybe

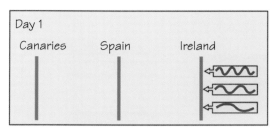

wave trains

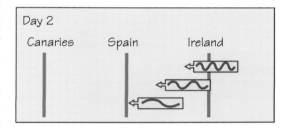

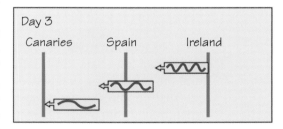

Figure 5.8
Schematic illustration of radial dispersion. Three different 'wave-trains', each of a different wavelength, set off at the same time from a low close to Ireland. As they travel away from the storm centre, the longer waves go faster than the shorter ones, progressively out-distancing them. By the time they arrive at the Canary Islands, the different wave-trains have completely separated out from each other.

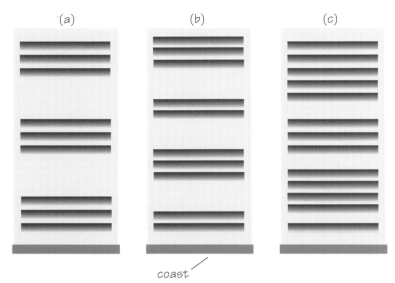

(a)   (b)   (c)

coast

Figure 5.9
Different combinations of 'settiness' in a swell arriving on the coast. Examples shown are: (a) sets of three waves at regular intervals with large gaps in between; (b) two swells arriving interspersed, one with two waves per set, the other with three waves per set; (c) a seemingly random mixture, due to either close proximity to the storm centre, or many swells interacting together.

ten waves, with much smaller waves, or nothing, in between. This is **grouping**. The particular manner in which waves are organized into sets is extremely important to us. For example, how many waves per set; how far apart the sets are, which wave in the set is the biggest – all these factors make some difference to a surf session. Different variations of **settiness** are suited to different surfing circumstances. If conditions are small and crowded, then the ideal situation is as many waves in a set, and as little time between sets as possible. However, this is exactly what you do not want if it is big. Ten waves per set spell disaster when it is huge, especially if you are likely to get caught inside. A nice, long but regular interval between sets is not a bad thing on big days, but can get frustrating if the waves are only about 1m (Figure 5.9).

Normally, the more dispersed the swell – in other words, the further you are from the storm centre – the more opportunity the group-forming mechanism has to work on the waves. Very close to the low pressure, sets are hardly noticeable. There may be bigger and smaller waves around, but the waves arrive more randomly and continuously. Thousands of kilometres away, there will be long periods of time when the ocean appears completely flat, and then the sudden arrival, seemingly out of nowhere, of a group of six waves, perfectly spaced and lined up.

Sometimes, even a long way from the storm centre, the sea looks mixed-up and random. This could be due to the superimposition of two or more swells arriving from different directions. They could have been produced by more than one storm, each having generated a swell destined to arrive on our beach at the same time. Or it could be due to different 'pulses' of swell having been sent by the same storm from two or more different places, while it was tracking across the ocean. If left to propagate further, these swells might pass through each other unheeded, each continuing to its own destination, or they might interfere in such a way as to produce a single, clean swell, with well-formed sets. There are many, many possibilities.

The fundamental mechanism that makes the waves form into sets is actually quite simple. It stems from the interference between waves of different wavelengths. But, of course, that is only a theoretical beginning to the story.

Waves bend as they propagate over different depths. Bells Beach, South Australia.

What is refraction?

Refraction is the bending of a wave as it propagates over different depths. When one part of a wave travels more slowly than another, the wave bends towards the slower part. A good analogy is travelling in your car with the brakes binding on one side: the car will tend to veer off to that side. With water waves travelling over varying bathymetry, any shallow areas of water will make the waves slow down. Therefore, waves will always be steered towards areas of shallow water. This is easy to visualize if you think of a single swell-line coming in towards a shallow reef next to a deep channel. The part of the swell over the reef slows down, whereas the part in the channel keeps going at the same speed, causing the wave to bend in towards the reef (Figure 6.1).

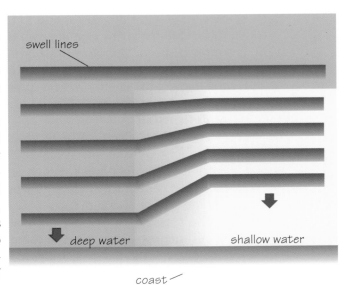

Figure 6.1 Schematic illustration of the simplest kind of refraction. The part of the wave front that propagates over shallow water slows down, and the part in deep water keeps going at the same speed. In the 'transition zone' between deep and shallow water, the wave front can be seen to bend towards the shallow part.

However, the sea-floor topography is never simple, and waves never come in from exactly the same angle, so there is an interesting array of different ways in which refraction can mould the characteristics of any and every surfing break. It can make the waves bigger or smaller; longer or shorter; faster; slower, or more hollow. It can also change dramatically the behaviour of the surf along a whole stretch of coastline, by acting on the continental shelf way before the waves are about to break.

Focusing and defocusing

Let us now consider a slightly more complicated situation. In (a) a shallow area with deep areas either side of it, or (b) a deep area with shallow areas either side of it, the way the waves refract when they hit this region is radically different.

Imagine a straight swell-line approaching a shallow slab of reef sticking out from the shore, with fairly deep water either side of it. The part of the swell-line that is over the reef will slow down, but the parts either side of it will keep going at the same speed. Therefore, the wave will bend inwards from both sides, towards the shallow water. This will concentrate all the energy into the middle, turning the whole thing into a peak. This kind of refraction is called **concave refraction** or **focusing**.

However, if the swell-line approaches a large, shallow area of water, with a deep trench in the middle of it, then the opposite will happen. The whole swell-line will slow down, due to the general shoaling of the water, except for the bit that is propagating over the deep trench. This will keep going at its original speed. Therefore, the wave will bend outwards, away from the deep water, and the middle part of the wave will become weaker and smaller. In the middle, the energy will be spread out over a wider area rather than concentrated into one spot. This kind of refraction we call **convex refraction** or **defocusing**.

The way in which waves break at a particular spot, and the configuration of the rock or sand which makes them break that way, defines that spot as a point-break, reef-break, beach-break, or whatever. By making the waves bend in a certain way before they break, refraction plays a vital role in deciding the nature of a surf spot.

The classic point-break, for example, has a set of characteristics all its own. Imagine a headland sticking out, with deep water off the end and shallow water to the side. There may be a reef running along the side of the headland, next to which could be a bay with a fairly flat beach. So the stretch of coast where the waves are going to break (on the side of the headland) is almost at right angles to the direction of wave-approach (Figure 6.2). As the swell-line comes in, one end of it will slow down as it hits the shallow water, and bend in towards the side of the headland, leaving the rest to continue on towards the beach. The waves will 'fan out' into the middle of the bay, having

been defocused by
convex refraction. The
energy will be spread
out over a wider area,
and the waves will be
reduced in size.

So, at a point-break,
the power and size
are reduced by defo-
cusing. But the waves
are much longer and
walled-up, the size of
the wave not dimin-
ishing as you go
down the line. In fact,
at some point-breaks,
the wave actually gets
bigger after take-off. Examples of where
you can see this happening are at J-Bay in
South Africa, Kirra in Australia, Tavarua
in Fiji, and Ala Moana Bowls in Hawaii

Another example of a classic kind of
break, with a completely different set of
characteristics, is the kind of reef-break
where the waves are focused on to a
slab of rock sticking out from a stretch of
open coast, or off the end of a headland.
The headland might be next to a deep
bay with a very steep beach, or even a
harbour, at its base. The reef is orientated so that the incoming swell-line hits
the end of the protruding reef first, the
rest of it continuing into deep water. The
swell-line will bend in towards the reef,
focusing all its energy over the reef and
leaving very little in the middle of the
bay (Figure 6.3).

The result is a bowling wave, where
all the energy that would have gone into
the deep area is concentrated into the
peak instead. This makes the wave bigger and more powerful. The reef acts as
a 'swell magnet', and the increase in size can sometimes
be surprising. At some breaks, the diminution of a long-

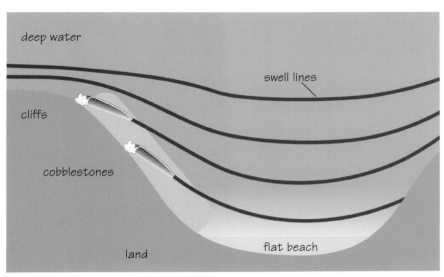

Figure 6.2 Convex refraction. A point with deep water
off the end, but shallow water in a bay next to it causes
the wave front to bow outwards, its energy being spread
over a wider area. The wave peels down the side of the

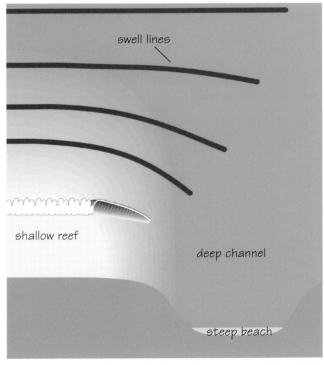

Figure 6.3 Concave refraction. A shallow reef
next to a deep channel causes the wave-front
to focus all its energy on to the reef area.

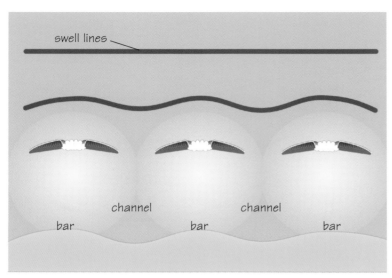

Figure 6.4 Typical fairly long beach with a series of sandbars along its length. A combination of concave and convex refraction produces a series of peaks with useful paddling channels in between. Without the sandbars, the approaching wave-front would just close out.

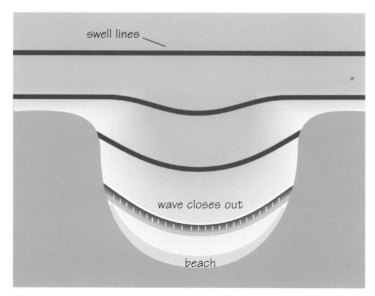

Figure 6.5 On a small 'pocket beach', refraction around the two headlands spreads the wave-front across the whole beach. The wave-front bows outwards, follows the depth contours, and eventually closes out. Not much good for surfing!

travelled swell due to circumferential dispersion is more than compensated for by the effects of local refraction.

A third kind of break is the beach-break. Here, refraction on to existing sandbars dictates where along the beach the waves will break, and how they will break. Most beaches, as long as they are of a certain length, contain some kind of along-shore undulations in the sand. Using the basic rules of refraction, we can see that the waves will be focused on to the shallow areas (the bars), and defocused away from the deep areas (the channels). This is the classic scenario of a beach-break, with various peaks along its length and channels in between (Figure 6.4).

Of course, not all beach-breaks are like this. For example, on a small 'pocket' beach there might be defocusing around both headlands, which spreads the wave out evenly along the whole length of the beach. Here you can see the characteristic 'bowing-out' of the wave before it eventually closes out across the whole bay (Figure 6.5).

Refraction is an essential ingredient in the chaotic recipe for the formation and maintenance of sandbars. The subject of how waves move sediment will be touched upon in Chapter 8.

Defocusing, at
Lynmouth, UK.

Focusing, at
Jardim do
Mar, Madeira.

Period dependency

One interesting feature of refraction is that it is wave period (and wavelength) dependent. The longer the period – the time between one wave and the next – the greater the refraction. In a new swell, where the long period waves arrive first, refraction has a more profound effect than it does later on in the life of the swell, when the shorter period waves arrive. This could mean that, when a new clean swell comes, any parts of the coast where there is more focusing, might show a lot more size than adjacent areas. If the waves are fairly short period – at the end of a swell, for example – then focusing won't have such a profound effect, and the waves might be the same size all along the coast. This is very useful for detecting whether there really is a new swell just starting, or it is just a figment of your imagination.

The reason why refraction is period-dependent is because longer waves produce larger water motions, which 'reach down' further beneath the surface. This means that longer waves feel the bottom, and hence start to slow down, before shorter ones. With uneven sea-bed contours, longer waves will therefore start to bend before shorter ones.

When a swell hits a point-break, it matters whether it is of long or short period. With a long-period swell, the waves bend around the point more, and so hook in and reel off down the line. If the period is short, the waves might end up going 'fat' into the middle of the bay, or have crumbling sections.

If the swell contains a multitude of different wavelengths at the same time, these will be separated out by period-dependent refraction. Refraction acts like dispersion, in that it allows waves of different wavelengths to adopt different propagation paths. This will be covered in more detail in Chapter 9, which looks at surfing in the fetch.

Summary

When waves come into shallow water they are affected by refraction. Because the sea-floor has many different forms, there are a great many ways in which refraction can form waves for surfing. The most basic concepts are of focusing and defocusing – concave or convex refraction – which either concentrates the wave energy towards one spot, or spreads it over a larger area.

When waves hit the edge of the continental shelf they always do so at some oblique angle. As they cross the shelf-edge, refraction makes them spread out over a wider area, and therefore lose height. The waves also propagate over the shelf more slowly, resulting in them having less power when they reach breaking point.

At a typical point-break, the waves are defocused, which means they lose some of their height, but to compensate for this they come in longer and more lined up. Also, the effect of refraction on a typical 'bowl'-type wave can easily be seen. Here, the swell is focused on to the reef, and all the power is concentrated in one spot. The wave is shorter, but more powerful, with a steeper take-off.

Refraction may affect a beach-break in many complicated ways. A typical scenario would be bars and troughs in the sand along the length of the beach, producing a series of alternate peaks and channels.

Refraction is period-dependent. In other words, waves with longer wavelengths bend more. This means that the long waves arriving on a stretch of coastline at the beginning of a swell, refract on to the reefs and points more than the shorter waves arriving at the end of the swell.

In Chapter 7, we will examine what happens when waves get into really shallow water and actually break. We will look into why some waves are grinding, thick-lipped monsters, while others are slow and mellow.

Box 4 Refraction: Snell's law

Snell's law enables us to calculate the angle of the refraction of a wave front when crossing a depth contour.

$$\frac{\sin(A)}{\text{speed in depth } h_1} = \frac{\sin(B)}{\text{speed in depth } h_2}$$

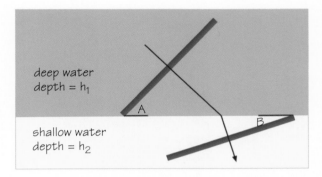

deep water
depth = h_1

A

B

shallow water
depth = h_2

In shallow water, the speed of a wave is proportional to the square root of the depth. Therefore:

$$\frac{\sin(A)}{\sqrt{h_1}} = \frac{\sin(B)}{\sqrt{h_2}}$$

Therefore, if we know the angle of approach of the wave front (A) and the two depths (h_1 and h_2), we can find out the resulting angle of the wave front after it has crossed the depth contour (B).

$$\sin(B) = \sqrt{\frac{h_2}{h_1}}\ \sin(A)$$

This simple example is the basis of much more complex refraction calculations.

For more information see:
Komar, P., 1998, *Beach Processes and Sedimentation* (2nd edn), Prentice Hall.

7 WAVE-BREAKING

Introduction

Wave-breaking is the beginning of the end of the journey. The Sun's energy, which momentarily organized itself into undulating perturbations of the water surface, eventually transfers itself, through wave-breaking, into the small-scale motions of turbulence and heat.

In this chapter we will look into what happens when waves break; why they break, and the shape of the waves when they break. We have already covered how different sea-floor topographies affect the bending of a wave-front just before it breaks, and how this can radically affect the final product – the ridden wave. Here we will look at the breaking of the wave itself, and the important differences between the ways in which the ocean finally unloads its energy on to the reef, point or sandbar.

Different kinds of breaking waves are suited to different styles of surfing. For example, mellow, slow breakers might suit the beginner or the laid-back longboarder who doesn't want anything too radical. These waves are easy to take off on, easy to ride, and easy to manoeuvre on, but might be considered boring by the more radical surfer. The fast, hollow wave, with the possibility of a tube, is more exciting, but at the same time, more risky and difficult to handle. And the inside-out, doubled up shore-break is heaven for boogie-boarders, but less practical for most stand-up surfers.

Each one of these kinds of waves has a different **breaking profile** – the shape of the wave looking side-on. The profile is dependent upon various factors, such as the bottom topography; the wind, and certain features of the waves themselves. All of these will be described in this chapter. We will start by looking into why waves break in the first place, and in what water depth, theoretically, they are supposed to break. Then we will describe how this depth can vary radically according to the above factors, and how this variation in breaking depth leads to differences in the profile of the wave. Make no mistake – the breaking wave profile is absolutely vital to anyone who is about to ride these waves.

Why do waves break?

In Chapter 6 we saw that, when waves start to propagate over shallow water, they generally slow down, and may tend to bend through the process of refraction. When they come into even shallower water, further changes take place, and eventually they break. But why do they break?

In shallow water, the speed of a wave depends only on the depth of the water. The shallower the water, the slower the wave. The exact relation between wave speed and water depth (namely, wave speed = 3.13 times the square root of the depth) was calculated by G.B. Airy in about 1845, together with a set of equations relating to water motion in and around a wave. This is called **Airy wave theory**, and can be used successfully when the height of the wave is considered negligible compared with the water depth.

However, to obtain more details about the motion of the wave when it comes into really shallow water, we must turn to a slightly more complex set of equations, devised by Sir George Stokes in 1880. **Stokes wave theory** considers not the whole wave, but the individual particles within the wave. While Airy wave theory took the wave height to be negligible, Stokes theory does not. Therefore, with Airy theory, the top of the wave and the bottom of the wave were virtually the same distance from the bed, whereas with Stokes theory the distance between the top of the wave and the bed is greater than the distance between the bottom of the wave and the bed.

Considering that wave speed – or, in this case, a particle within the wave – is directly related to water depth, we can see that the further from the bed a particle is, the more water it has underneath it, and hence the faster it travels. Therefore, the top of a wave travels faster than the bottom. The action of the bed is felt more at the bottom of the wave than the top, and therefore slows down the bottom more than the top. This effect is progressively more

A time sequence of a breaking plunging wave.

pronounced as the water gets shallower, so a point is reached when the top of the wave overtakes the bottom, making it spill forward and break. The wave is 'tripped up' – just as you might be tripped up by someone as you walk along – making the bottom half of you travel more slowly than the top, so that you fall over head first.

When will a wave break?

How can you tell when a wave will actually break? In other words, at what depth of water does a typical wave break? Through years of observations, experts have come up with an average depth of about 1.3 times the height of the wave. That means that a typical 1m wave will break in about 1.3m of water. Very useful if you want to know how much water there is under you when you take off on a wave; or, perhaps, if you want to design an artificial reef.

However, we all know that the depth under a breaking wave can vary enormously, and it is this varying depth that mainly governs the wave profile. At Pipeline, for example, you might get 3m waves breaking in less than 1m of water. Or on a very flat beach on an onshore day, you might get 1m waves breaking in 3m of water.

One major factor that the breaking depth – and hence the wave profile – depends upon, is the suddenness with which the bottom topography changes as the wave propagates towards the shore. If the transition from deep to shallow water is very sudden – for example, over an offshore coral reef – then, momentarily, the wave is inevitably over very shallow water without having broken. At this point, the height of the wave is quite significant compared with the depth of water, and the distance between the top of the wave and the sea-floor is much more than the distance between the bottom of the wave and the sea-floor. Remembering that wave speed is dependent on depth, you can see that the speed at the top of the wave will be much greater than that at the bottom of the wave. This will cause the wave to throw out as it breaks, making it steep, hollow and fast.

on its way to somewhere. This **velocity skewness** is the key to sediment transport, and the starting point for the formation of sandbars (Box 6).

Of course, the sediment does not just move onshore and offshore. Sandbars are three-dimensional features, and they depend upon the sediment moving along the shore as well. Along-shore sediment transport is a secondary consequence of the on-offshore transport described above, but the driving force is still the action of the waves themselves.

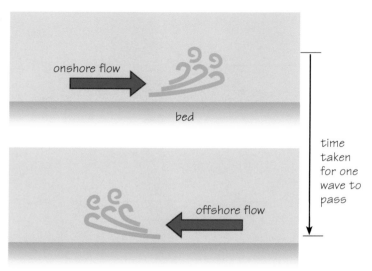

Figure 8.2
Sediment is 'scooped up' by the backward and forward motions of the water near the sea-bed.

The summer-winter profile

Before going into more detail about the formation of sandbars, we will look at a 'two-dimensional' example of a change in beach topography from the action of waves. The **summer-winter profile** is one of the simplest examples of how the on-offshore movement of sediment changes the shape of the beach from one season to another.

Because the flow of water dragging on the sea-bed might be greater in one direction than the other (wave skewness, mentioned above), sediment generally tends to accumulate right under the place where the waves are breaking. This is because the skewness causes an average transport of sediment in an onshore direction outside the break point, and an average transport of sediment in an offshore direction inside the break point.

According to this simple rule, a sandbar will form more or less under where the waves are breaking. In winter, when the surf is big and stormy, the waves tend to break further out than in summer, when it is small and clean. So the typical winter profile of a beach might consist of a sandbar maybe 200m offshore, and the summer profile might have a steep bank very near to the shoreline. This is often the case, where, in winter, a bar forms out the back causing waves to break and then back off. In summer, this bar is less likely to form and, quite often, the beach-breaks are more 'in control', with good quality small waves breaking close to the shore (Figure 8.3).

An interesting twist on the summer-winter profile is the dire consequences that a sudden large swell at the end of summer can have on beach stability. In Europe, the occasional ex-hurricane redeepens to form a powerful mid-latitude depression, which might bring five-metre waves after three months with nothing over a metre.

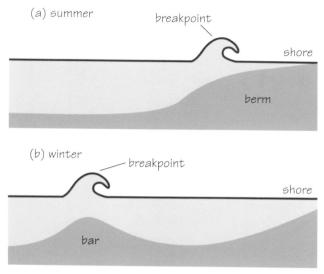

(a) summer

breakpoint

shore

berm

(b) winter

breakpoint

shore

bar

Figure 8.3 The summer-winter beach profile. In summer the beach is bigger; there is a **berm** – a steep bank of sand formed at the shoreline – at the top of the beach, but no offshore bar. In winter, when the waves break further out, all the sand moves offshore to make a bar.

However, if there is a sudden storm after a long period of summer calm, then the beach does not have the luxury of an offshore bar to protect it. The feedback loop is broken. The natural coastal protection mechanism does not work any more, and the big waves will come right in, devouring the beach, its occupants and their property.

The winter profile normally remains as it is for several months through a feedback mechanism between the waves and the sandbar itself. The waves are generally large, so the bar begins to form out the back under where they break. Any further swells tend to break on the bar because of the shallow water. This causes further accumulation of sediment, and so reinforces the bar. Therefore, during winter, the very fact that the waves break out the back on the bar means that they will not come in and break on the shore and start eroding away the coast. They will dissipate all their energy on the bar.

The formation of sandbars for surfing

We now have an idea of how sediment moves in an onshore-offshore direction to produce a sandbar. But that does not tell us why, at some beaches, there are a series of sandbars along the length of the beach producing good, reliable peaks, rights and lefts, and channels either side, whereas at other beaches this never quite happens.

The sandbar described above in the winter profile is not just a great long sausage that stretches the whole length of the beach. It is more like a series of lumps and dips. The peaks are where the lumps are, and the channels where the dips are. But how does the bar get into this shape in the first place?

The answer lies in a chaotic system, where imperceptibly small changes in input produce vastly differing outputs. For example, if the sandbar is just slightly higher in one place than another, then any incoming waves are focused directly on to that spot by refraction. This makes them bigger, and more likely to break, not only reinforcing the bar by the basic sediment transport mechanisms mentioned above, but also by causing water to run off the side of the bar into a deep-water area. This water then starts flowing out to sea again in the form of a rip, which tends to gouge out the channel even more. Therefore, the difference in the height of the sand between the lump and the dip increases, which causes the waves to be focused even more on to the lump. The increase in wave-breaking causes more sediment to accumulate under that spot, and a greater flow of water into the channel.

This feedback allows the system to reach a state of equilibrium, forming the classic beach-break with a series of peaks all along its length. In effect, the sandbars organize themselves. It is a **self-organizing system**.

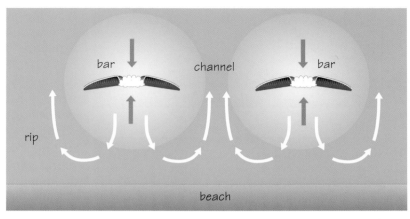

Figure 8.4 In the sandbar-channel system (previously seen in Figure 6.4), sandbars are kept in place by waves breaking over them and rips running in between. Dark arrows indicate sediment movement; light arrows, water movement.

If this system works properly, then we will have a perfect set-up for surfing: rights and lefts, always breaking in the same spot; a choice of peaks so there is plenty of room for everyone, and convenient paddling channels in between each peak (Figure 8.4). However, very often the situation is not quite perfect. Some other factor may inhibit the self-organization of the system. For example, there might be a large **tidal range** that does not give the sandbars enough time to set themselves up; or extreme refraction around two headlands either side of a restrictively small beach, causing the wave always to close out. The beach might work well only at certain swell sizes; or the wave conditions required to set up a good series of sandbars might occur only a few times a year.

Oblique wave approach

It is obvious that waves do not always come in exactly perpendicular to the coast. Small variations in swell direction can set off the chaotic chain described above, whereby the existing sandbar begins to become uneven in the along-shore direction. But how is the beach topography affected if the swell comes in at quite an oblique angle?

Let us think of a north-facing beach, with the swell coming in from either the north-east or north-west. Bearing in mind that sand tends to accumulate under the break-point, it is apparent that a bar will form under where the waves are breaking. This bar will be orientated in the direction of the waves. For example, if the swell is from the north-west and remains that way for some time, then a series of sausage-shaped sandbars will form, orientated parallel to the swell-lines. If the swell continues to come from the same direction, then the waves will probably just close out on these bars.

However, on some beaches, the swell direction may be from one direction for some time, and then radically change to another direction. If, on our north-facing beach, the swell suddenly switches to north-east then, at

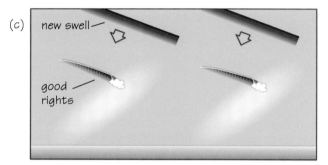

Figure 8.5 If waves approach a beach at some oblique angle (a), then sandbars will eventually form at an oblique angle (b), causing the waves to close out. A sudden change in swell direction (c) will result in good peeling waves, for a while at least.

first, the swell lines will come in at a very oblique angle to the 'north-west' sandbars. This could make the waves break really well, peeling along from one end of the bar to the other. For a while the beach will have a series of perfect rights. Eventually, the continuous north-east swell will cause the bars to orientate themselves the other way round, and the waves will close out again (Figure 8.5).

This example is an idealized and simplified scenario, but at times such situations do occur and, when they do, it is interesting to know why. Obviously, there are many more complications, which we will not go into here. For example, what happens to the sand when two swell directions come in at once? And what happens when various combinations of the recent history of wave action occur – a large storm from the north-west, followed by two days of big, clean northerly swell, followed by three days of small, choppy waves from the north-east, perhaps?

The study of coastal geomorphology

The subject of wave-driven sediment transport is vast, and is concerned with far more than the formation of sandbars that may or may not be good for surfing. The biggest motivation for scientists to study sediment transport is for coastal geomorphology – how the topography of the beach changes as energy is input to it from the breaking waves. The way the coastline changes over long time-scales is extremely important to all of us. The beach is our working environment, the very substrate upon which we perform our art. With phenomena such as global warming and sea-level rise, who knows if, in 20 years time, all our beaches might get washed away? So this is the motivation: to one day be able to predict quantitatively how the coast might change according to different wave conditions. This is not to say that we should necessarily interfere with nature, or do anything to try to prevent it.

Summary

In this chapter we have looked briefly at a few ways in which the action of waves on the sand can alter the way the waves break, making them either good or bad for surfing. The small amount of the Sun's energy that was fleetingly manifest in wind, causing a series of undulations on the surface of the ocean, called waves, is partly dissipated with the breaking of the waves themselves. But it is also partly transferred into altering the shape of the beach. This stage is at the end of the recognizable life of the ocean wave: it has dissipated much of its energy into heat and turbulence from breaking, and the remainder becomes the input to the system of coastal geomorphology. It is a complicated world of chaos and feedback.

The formation of good sandbars for surfing depends on a complex system of energy feedback: the breaking waves modify the beach topography, and the topography in turn modifies the way the waves break. In general, beaches are highly dynamic and their shape can change rapidly. However, some spots, such as those near river mouths and piers, have a degree of built-in stability.

The basic mechanism that allows the waves to move the sediment is the near-bed water motion under the waves. The water drags along the bottom, transporting the sediment. If the motion is greater in one direction than the other, then the sediment will have a net displacement.

A simple example of onshore-offshore sediment transport is the summer-winter profile, where a sandbar is formed under the position where the waves break. It will be further offshore in the winter and nearer the shore in the summer. If a large storm hits at the end of the summer, the beach will not have the offshore bar to protect it – so beachside property is likely to be damaged.

The classic bar-trough system, with a series of good sandbars along the length of a beach, and paddling channels in between, is a consequence of a chain of events that starts with slight differences in the existing topography. Feedback between the waves breaking over the bars; the accumulation of sediment below the breaking waves, and the gouging out of a channel by the rips, maintains these bars in place. Obviously, not all beaches are like this, for a number of reasons, including tidal movements and inconsistent swell sizes.

Coastal geomorphology is studied by scientists to try to predict how wave action changes the shape of our coasts. This is very important, especially in view of recent concerns about sea-level rise and increasing wave heights from global warming.

Box 6 Velocity skewness

Graphical illustration of velocity skewness. As a wave passes overhead on the surface, the water near the bed first goes fast in an onshore direction for a relatively short time, and then goes in an offshore direction more slowly, but for a relatively long time. The onshore movement is fast enough to scoop up sediment, but the offshore movement is not.

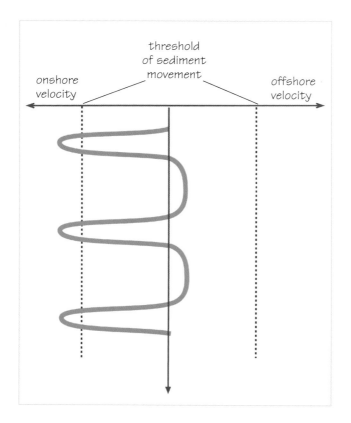

behave differently when they heat up and cool down. Land heats up and cools down very quickly; sea takes much longer to heat up and cool down. On a typical stretch of coastline, the sea might change its temperature only a few degrees throughout the day, while the temperature of the land might vary by tens of degrees.

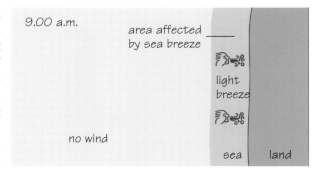

This difference in daily temperature variation is due to the fact that water has a higher specific heat capacity than land. This means that much more energy must be put into water than land to raise its temperature by the same amount. Therefore, water takes much longer to heat up than land. But once water is warm, all that energy is locked up inside it.

So what happens in the cooling process? Water, with its great bulk of stored energy, is slow to lower its temperature, because it has to get rid of all this energy. Land, however, having accumulated a lot less energy in the first place, takes a fraction of the time to cool down.

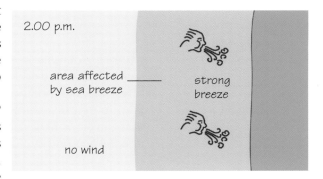

In the morning, before the Sun gets up into the sky, the land may be slightly colder than the sea, or they may be both around the same temperature. Throughout the morning, the Sun's energy heats up both land and sea, but because of its lower specific heat capacity, it heats up the land much more than the sea. By midday, the land is much warmer than the sea.

The air that is over the land is relatively warm, and so it starts to rise, due to the process of convection. The air over the sea tends not to rise, because it is not so warm. The air that rises over the land leaves a 'gap', which must be filled by something. The nearest available substance to fill this gap is the cooler air lying over the sea, which starts to generate horizontal movement of surface air – called **advection** – from the sea to the land: in other words, an onshore breeze. As long as the land is warmer than the sea, this kind of circulation pattern continues.

Figure 10.2 The sea breeze has a limited fetch that does not extend very far offshore. However, during the day, this fetch increases, as does the strength of the breeze. The angle of the breeze also tends to change throughout the day. A Northern hemisphere example is shown, which turns to the right.

Later on into the evening, the land begins to cool off, and the sea breeze dies right down, leaving glassy conditions. The air above the land has stopped rising, and may even begin to start sinking down again. If the land temperature gets down below that of the sea, the air drifts back from the land to the sea, producing a light offshore breeze. The land rarely gets much colder than the sea, so the land breeze is almost never as strong as its onshore counterpart. Note that the land breeze can be slightly stronger in areas where mountains border the coastline. This land breeze continues as

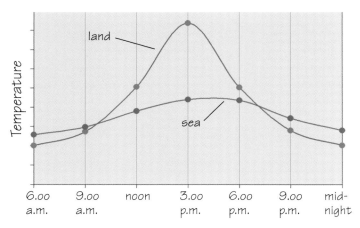

Figure 10.3 Graph showing how land and sea temperatures vary throughout the day. Clearly, the land temperature varies a lot more than the sea, and the sea takes longer to heat up.

long as the sea is warmer than the land, right on into the next morning if we are lucky. Then, as the Sun starts to heat things up again, the cycle begins to repeat itself (Figures 10.3 and 10.4).

Where and when does the sea breeze happen?

The principal requirement for a sea breeze is for the location to be generally hot and sunny. In latitudes where the climate is relatively cold, or it is cloudy most of the time, a sea breeze rarely occurs. The characteristics of the land and sea also affect the likelihood of a sea breeze. For example, if the land has a feature that gives it a particularly low specific heat capacity, or one that makes it heat up and cool down even faster, this will encourage the sea breeze. And if the sea has some characteristic that helps to keep it the same temperature, this will also help the sea breeze to form.

The coastal deserts on the western side of continents are places where conditions are perfectly suited to a sea breeze. The land has almost no vegetation, which allows it to heat up and cool down extremely quickly. These places tend to have coastal upwelling and cold surface currents, so the water is being replenished constantly, and has little chance to change its temperature. Because of this constant supply of cold water, and the ability of the land temperature to rise many tens of degrees during the day, the sea breeze can sometimes blow quite strong.

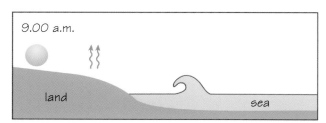

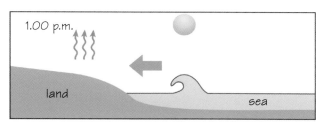

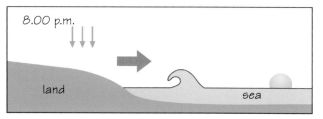

Figure 10.4 At about 9.00 a.m. the land is just starting to be heated by the Sun. Soon after midday, it has heated up quite a lot, and the air above it has started to rise. Cooler air from the sea rushes in to replace it. At about 8.00 p.m. the temperature of the land has cooled to just below that of the sea, allowing air to sink down over the land, and squeeze the warm surface air out over the sea.

Conditions change from early morning offshores (above), to middday mush (below), to evening glass-off (overleaf).

Evening glass-off.

If you live somewhere where the sea breeze is strictly a summer occur-rence, at the beginning of the summer those nasty onshores will probably pester you almost every day. As the months wear on, they get less and less frequent until the autumn, when they more or less disappear altogether. At the same time, the offshores will tend to last longer and longer into the morn-ing. This is because the average sea temperature is getting progressively higher all through the summer, but the land temperature is swinging up and down on a daily basis. So, at the beginning of the summer, during a typical sunny day, the land gets much warmer than the sea, whereas at the end of the summer, it might get only a few degrees warmer. In other words, during the months when the air temperature gets well above the sea temperature during the day, you are likely to get a sea breeze; in the months when the air temperature is below that of the sea, a sea breeze is much less likely (Figure 10.5).

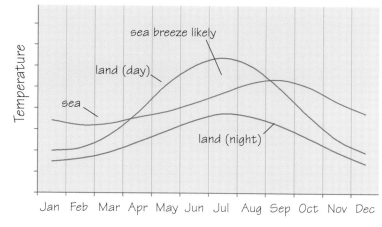

Figure 10.5 Typical land and sea temperatures throughout the year for some arbitrary place in the Northern hemisphere. A sea breeze is more likely between about April and August, when the land is warmer than the sea during the day.

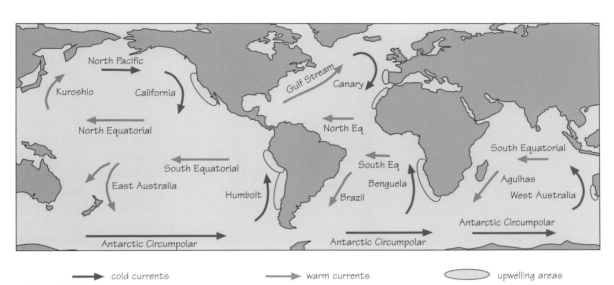

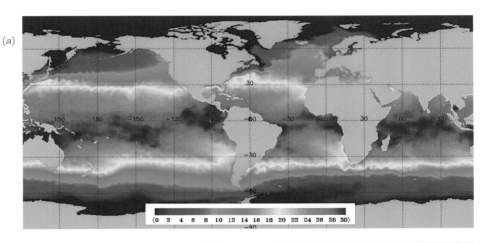

(a)

Figure 11.1
Simplified map
of world ocean
currents showing
upwelling areas.

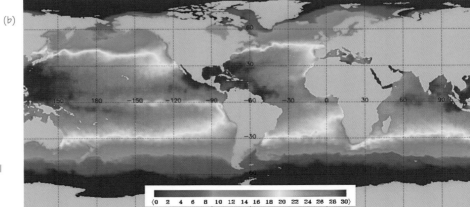

(b)

Figure 11.2
Typical average global
sea temperatures for
(a) February and
(b) August.

Plan view

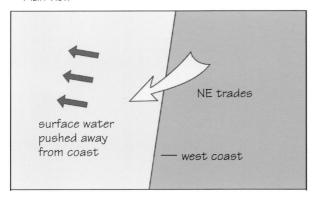

Side view

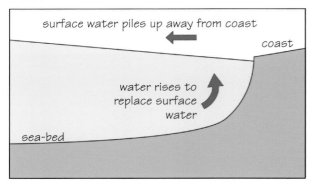

Figure 11.3 Plan view and side view of upwelling mechanism.

However, places well away from the reach of ocean currents are like stagnant pools. The surface water hardly moves, so it really does have a chance to heat up and cool down with the seasons. In the shallow waters of the southern North Sea, for example, where the Atlantic currents cannot reach, the temperature goes up and down like a yo-yo. The combination of being situated close to a large land mass, away from any surface water supply, and being relatively shallow, ensures a large summer/winter temperature swing. Somewhere completely surrounded by land, such as the Great Lakes, is not influenced at all by ocean currents, and the water is almost totally at the mercy of local heating and cooling.

The slow heating up and cooling down of the sea

Another strange behavioural pattern of our oceans is that, because water has such a high specific heat capacity, it takes a very long time to heat up and cool down (this was mentioned in Chapter 10). If you live somewhere where the water temperature varies quite a bit throughout the year, you might find the seasonal variations to be totally out of phase with the seasons themselves. Maximum and minimum sea temperatures tend not to occur in mid-summer and mid-winter – in fact, they occur months afterwards. In early spring, for example, just when the weather is starting to warm up, the sea might still be cooling down. And in autumn, just when the air is starting to become a little chilly, the water is at its warmest, or may still be warming up.

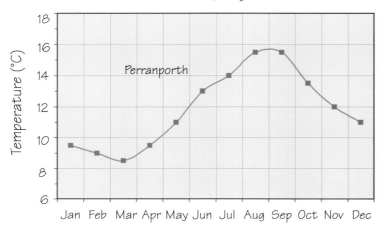

Figure 11.4 Average sea-temperature data for Perranporth beach, UK. The water temperature is at its minimum in March, and its maximum around the end of August. The sea takes about two months to warm up and cool down.

The oceans require a massive energy input to allow their temperature to rise just a few degrees, which may take months. Then, when the heat input is reduced, it takes months for the oceans to cool down again (Figure 11.4).

Summary

In this short chapter we have looked at how and why the water temperature varies throughout our surfing waters in a way that, at first sight, is not obvious. This could all be worth thinking about if you are planning a trip somewhere uncharted. Once you have found out the prospects of surf, do some research into the water temperature. This might save you (a) carrying extra boots, gloves and helmet for nothing; (b) coming back with a great suntan, but having surfed for only 20 minutes a day because your 2mm spring suit was not enough, or (c) coming back having been unable to surf at all, because you were there in the 'frozen' season.

The water temperature at a particular spot often bears no relation to the latitude, or the seasons. In Europe, for example, there are many places where the variation of water temperature with latitude is the opposite from what one would expect. One reason for this is the ocean currents, which constantly replace the surface water, not giving it a chance to warm up and cool down locally. Many mid-ocean spots have a tiny seasonal temperature variation, whereas places situated away from the influence of any external water supply have much larger variations throughout the year.

If the temperature does vary seasonally, then the situation is still not totally straightforward. Because of the large specific heat capacity of water, the annual temperature cycle of the sea may be totally out of phase with the seasons themselves.

Box 8 Specific heat capacity

The *specific heat capacity* of a substance is the amount of energy required to raise one kilogram of that substance one degree centigrade. To find out how long it takes to heat up a substance one degree, we can use:

$$\boxed{\text{time taken}} = \boxed{\text{mass}} \ \text{X} \ \boxed{\frac{\text{specific heat capacity}}{\text{rate of heat input}}}$$

To compare how long it takes to heat up the sea compared with the land, we can do a simple example. The specific heat capacity of water is 4,187 joules per kilogram per degree centigrade (J/kg/°C), and the average specific heat capacity of sand, brick and stone (typical substances found near the coast) is 825 J/kg/°C. If the rate of heat input is, say, 1,000 joules per second, then to heat 1,000 kg of seawater 1°C takes:

$$1,000 \ \text{X} \ \frac{4,187}{1,000} \ = \ 4,187\text{s} \ = \ \underline{1 \text{ hour } 10 \text{ mins}}$$

But to heat 1,000kg of 'land' 1°C takes only:

$$1,000 \ \text{X} \ \frac{825}{1,000} \ = \ 825\text{s} \ = \ \underline{14 \text{ mins}}$$

So we can see that, even without the extra regulatory effects of ocean currents and upwelling, the sea takes much longer to heat up and cool down than the land.

For more information see:
Duxbury, A.B., Duxbury, A.C. and Sverdrup, K., 2000, *An Introduction to the World's Oceans* (6th edn), McGraw Hill.

The information in this chapter should help you to choose your destination for a surf trip. Having drawn up a short list of the best possible surfing areas according to the time of year, if you are able to wait till about three days before leaving, before making the final decision, you will also be able to benefit from short-term forecasting techniques. These will be discussed in Chapter 14.

Wave measurement from space

To gain information about the average height of waves over large areas of the ocean, the best thing we can use is a satellite. A satellite can measure wave characteristics in several ways, the simplest being **radar altimetry**, where the satellite emits a pulse of radiation towards the Earth about once every thousandth of a second. The signal is reflected off the sea surface and returns towards the satellite, where it is detected by a sensor. As with a typical echo-sounder on a boat that measures the distance between the boat and the sea-floor, the time taken for each pulse to return is used to calculate the distance between the satellite and the sea surface. This is a relatively crude technique, which cannot distinguish very well between long- and short-period waves, and whose accuracy is not yet particularly good. However, it is a very good way of estimating wave heights over the whole Earth. This would be impossible using conventional *in-situ* equipment such as wavebuoys.

The enormous amount of data received by a wave-measuring satellite circulating around the Earth can eventually be processed into charts containing contours of average wave heights for each month or week, or any other period, throughout the year. If we required monthly averages, we would need 12 world maps – one for each month – each with different areas signifying different average wave heights for that particular month.

To achieve this, first the surface of the Earth would be divided into an arbitrary grid, containing a number of points at, say, 100km intervals. The satellite would already have been busy gathering wave heights for every point on this grid for a long time – preferably several years. Then, to obtain the January average wave heights for just one grid point, all the heights measured in January would be averaged. This would be repeated for all the other grid points to get the January heights for the whole globe. To complete the annual picture, the whole process would be repeated for the other months of the year.

World wave heights for January and December

We will look at two of these charts – obtained using the technique described above – for January (Northern winter, Southern summer) and July (Northern summer, Southern winter). These will give us an idea of typical summer and winter wave heights for both hemispheres (Figure 13.1).

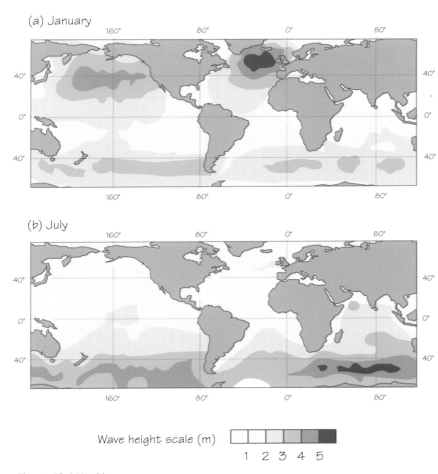

(a) January

(b) July

Wave height scale (m)

1 2 3 4 5

Figure 13.1 World wave heights for (a) January and (b) July.

When looking at these charts we must realize that the satellite cannot distinguish very well between storm waves and swell waves. Therefore, the main use of the charts is to show us where the storm centres are most likely to be (although waves that have propagated away from the storm centre can also be detected, but not identified as such – see below). So we must look at the contours of large wave heights and relate them to where the storm centres are likely to be, and how intense they are likely to get.

Having a series of charts for different times of the year is much better than the all-year-round average wave height often quoted for many spots. The latter is of very limited use, as it does not give us any idea of seasonal variation. For example, a 1m year-round average could mean 1m every day, or it could mean 36.5m for ten days of the year, and then flat for the rest of the time. With the information we have here, we can estimate how consistent an area is, as well as how big it will get.

General observations from the charts

What can these charts tell us? First, comparing the Northern hemisphere with the Southern, we can see that overall wave heights in the Northern hemisphere vary throughout the year more than they do in the Southern hemisphere. In other words, on average, the North is very big in January and very small in July, whereas the South is quite big in July and quite small in January (Figure 13.2).

This is an important point, as it shows why some spots on the west and south-west coasts of Australia, South America and Africa are very rarely flat.

Some Northern-hemisphere spots, however, can be huge in the winter and virtually flat all summer. At Waimea Bay, for example, you might find boats anchored in the bay during the summer, and you would never know that, during the winter, the place can turn into a cauldron of whitewater with ten-metre closeouts and huge rips.

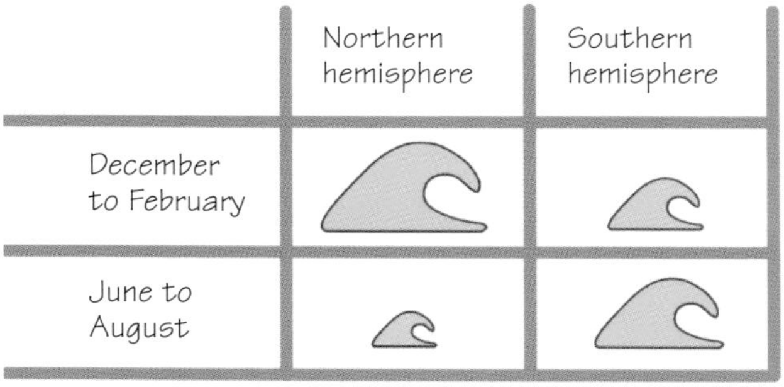

	Northern hemisphere	Southern hemisphere
December to February		
June to August		

Figure 13.2 Schematic representation of seasonal variation of wave heights. In the Northern hemisphere there is a greater difference between summer and winter. Also, in theory, there is 'more surf' around the world between December and February.

The reason for this is the difference between the large-scale circulation patterns in the Northern and Southern hemispheres, mentioned in Chapter 2. The roaring forties continue to blow around the Southern hemisphere, shifting north and intensifying during the winter, but still having some influence in the summer. This means that the waves in winter will be bigger, but there will still be waves generated during the summer.

The other interesting thing that the charts tell us is the difference between wave height in each ocean. In January, the North Atlantic appears to have the edge on the Pacific for shear wave height. However, this does not necessarily mean that the coasts of Europe will have larger rideable waves than those of western North America and Hawaii, since the North Atlantic has less space to allow the waves to propagate away from the storm centre.

In the oceans of the Southern hemisphere, the area with the biggest waves appears to be the Indian Ocean. This implies that the west-facing coasts of Australia, for example, receive slightly bigger waves than places at the same latitudes in southern Africa or South America.

Surfability and local factors

Note that wave heights measured by satellite include, embedded within them, waves generated from various different sources. The satellite measures 'raw' wave heights: it can neither measure the direction of wave propagation, nor distinguish between waves generated locally and waves that have propagated in from elsewhere.

Satellite charts show no information about different sources of swell, and no information about the quality of waves when they reach the coast. The waves may arrive super clean and lined-up, or totally out of control, and local winds may enhance surfing conditions, or worsen them.

Therefore, to complement the general information provided by the satellite, we will look at more specific details about the sources of surf in key areas around the world. For each area, we show a summary of the principal swell sources; their approximate direction and intensity, and at what time of the year we should expect swell from each source.

Wave climate: zone by zone

- *Western Europe:* Main swell source is mid-latitude depressions that track across the Atlantic; strongest from December to March. Ex-hurricanes that redeepen into mid-latitude depressions can produce surf on south- and west-facing coasts, August and September. Small, Southern hemisphere swells can be detected, but are almost never big enough to surf. Ireland and the UK receive the most surf, but are plagued by windy or choppy conditions and the weakening effect of the continental shelf. France, Spain and Portugal are best positioned to handle big swells.
- *The Mediterranean:* Surf is produced mainly by local windseas from depressions that enter the Mediterranean from the Atlantic, December to February. On the west side, the *poniente* – a strong westerly wind that funnels through the Straits of Gibraltar – may act as a small 'wave-engine', producing real, but small-scale groundswells that propagate towards Sardinia, Corsica, the Balearics and Sicily.
- *North-West Africa:* Receives the same swells as western Europe, but may be smaller and more lined up due to greater propagation distances. Very strong northerly-tracking lows produce much cleaner surf than the more southerly-tracking weaker depressions. Trade-winds can produce some surf in the Canary Islands, May to August. Some swell from hurricanes, August to October.
- *West Africa:* South-facing coasts in the Gulf of Guinea receive south swells from the roaring forties, but these arrive very weak. Note that there are no hurricanes in the South Atlantic.
- *South-West Africa:* Year-round, consistent swells from the roaring forties, biggest June to August; but may be out of control on southernmost beaches with west or south-west orientations.
- *South-East Africa:* South swells from the roaring forties, best June to August. Also receives surf from cyclones off Madagascar, November to February. Local windseas allow year-round rideable conditions for the beaches of Durban and the south-east of South Africa.
- *Western Indian Ocean:* Cyclones north of the Equator, May to June and October to November, bring surf to the Arabian Sea and Bay of Bengal coasts. Cyclones south of the Equator, November to February, enhance the surf of Madagascar, Mauritius and Réunion, which are also better positioned to receive south and south-west swells from the roaring forties. These swells arrive rather weak and short-lived on the more easterly-facing coasts; best June to August.
- *Eastern Indian Ocean:* This is the best area to receive Southern Ocean swells. Indonesia is far enough away from the roaring forties for swells to be clean and wind conditions to be favourable, but still gets a constant supply of reasonable-sized waves; best April to September.

pressure reduces by more than 24hPa in 24 hours.

extra-tropical cyclone, 24. *An atmospheric disturbance found at latitudes between about 30° and 70°, resulting in a large mass of cyclonically rotating surface air (clockwise in the Southern hemisphere and anti-clockwise in the Northern hemi-sphere). The most important feature for producing surf. Same as* **low pressure** *or* **mid-latitude depression**.

F

FAS, *see* **fully arisen sea**.

fetch, 12. *The length of an area of sea surface over which the wind blows continuously to generate waves.*

fetch-limited, 35. *In* **SMB theory**, *when the wave growth is limited by the fetch – i.e., the wave height depends only on the wind speed and the fetch.*

focusing, 52. *The bending inwards of a wave-front, due to the middle propagating more slowly than the end(s). The energy will be concentrated into a smaller area, therefore the wave will get bigger, but become shorter. The same as* **concave refraction**.

form number, 111. *A number used to quantify whereabouts the tidal regime is between* **diurnal** *and* **semi-diurnal**.

fully arisen sea (**FAS**), 35. *In* **SMB theory**, *when the waves cannot grow any higher for a particular wind speed, no matter how long it blows, or how big the* **fetch** *is. In an FAS, the*

wave height depends only on the wind speed.

fully developed sea, 79, see **fully arisen sea**.

G

gravity waves, 34. *Sea-surface waves that have gravity as their restoring force.*

groundswell, 63. *Waves that are propagating away from the storm centre 'on their own', without any further input of energy from the wind.*

group speed, 42. *The speed at which the energy of a swell is carried. In deep water the group speed is half the individual wave speed.*

groupiness, 10. *The characteristics of the waves in terms of how they are grouped – e.g., the number, and variability of the number, of waves in the group, and the time between the arrival of successive groups. Same as* **settiness**.

grouping, 45. *The formation of waves into sets or groups, due to the interaction between two or many different wave trains consisting of waves of different periods and directions.*

H

hurricane, 118. *The name give to a* **tropical revolving storm** *(TRS) in the North Atlantic and Eastern Pacific Ocean, particularly around the Caribbean and Central America.*

I

interpolation, 108. *The estimation of the value of some parameter at a point, based on its value at two or more sur-*

rounding points. Useful for plotting contour charts.

Iribarren number, 67. *A number used to distinguish between different kinds of breaking waves. Calculated from wave height, wave period and beach slope.*

isobar, 27. *A line of equal pressure.*

J

jet stream, 28. *A strong circumpolar westerly air stream at mid-latitudes, at heights of between 5km and 10km. Highly influential in the formation and movement of low pressure systems.*

Joint North Sea Wave Experiment (**JONSWAP**), 36.

K

knot, 123. *Traditional measure of speed at sea, equivalent to one nautical mile per hour or 0.51m/s.*

L

low pressure, 24. *An atmospheric disturbance found at latitudes between about 30° and 70°, resulting in a large mass of cyclonically rotating surface air (clockwise in the Southern hemisphere and anticlockwise in the Northern hemisphere). The most important feature for producing surf. Same as* **mid-latitude depression** *and* **extra-tropical cyclone**.

M

macrotidal, 111. *Where the tidal range is more than 4m.*

mesotidal, 111. *Where the tidal range is 2–4m.*

microtidal, 111. *Where the tidal range is less than 2m.*

mid-latitude depression, 24. *An atmospheric disturbance found at latitudes between about 30° and 70°, resulting in a large mass of cyclonically rotating surface air (clockwise in the Southern hemisphere and anticlockwise in the Northern hemisphere). The most important feature for producing surf. Same as* **low pressure** *and* **extra-tropical cyclone**.

monsoon, 93. *Name for seasonal winds. First applied to the winds over the Arabian Sea, which blow for six months from the north-east and for six months from the south-west; now extended to similar winds in other parts of the world. In India, the term is popularly applied to the south-west monsoon and also the rains that it brings..*

N

non-linear transfer, 36. *The transfer of energy from waves of one period to another during the propagation of a swell. Can be important in wave prediction models.*

North Atlantic Oscillation (NAO), 131. *A cyclic climatological pattern relating to the relative strengths of the Icelandic low and Azores high.*

O

obliquity of the ecliptic, 18. *The angle of 23.5° between the axis of the Earth's rotation around itself and the axis of the Earth's rotation around the Sun.*

occluded front, 27. *When a low pressure system is weakening,* the warm and cold fronts join together, forming an occluded front.

opposition, 104. *When the Moon and Sun are lying on opposite sides of the Earth, resulting in a full Moon and spring tides.*

P

parameterization, 36. *In wave prediction models, the technique of 'forcing' the spectrum to a prescribed shape, without having to calculate the full spectral characteristics every time.*

peak period, 128. *The wave period at which there is most energy.*

period, see **wave period**.

phase speed, 42. *The speed of each individual wave.*

plunging breaker, 64. *The technical term for a tubing wave with a reasonable amount of power.*

polar front, 25. *An imaginary line at mid-latitudes where cold Equator-ward air, originating from high latitudes, meets warm poleward air from low latitudes. The polar front is an example of a place where low pressures are likely to form.*

poniente, 78. *The strong westerly wind that sometimes blows through the Straits of Gibraltar.*

pressure gradient, 122. *The change in pressure over a given distance. The stronger the pressure gradient, the stronger the wind.*

Q

quadrature, 104. *When the Moon and Sun are lying at right-angles to each other relative to the Earth, resulting in a half Moon* and neap tides.

R

radar altimetry, 113. *Technique of using radar, usually mounted on a satellite, to determine the height of the sea surface. Can be used to obtain wave heights over large areas.*

radial dispersion, 43. *In a propagating swell, because longer waves travel faster than shorter ones, the longer ones will progressively out-distance the shorter ones.*

radiative transfer equation, 35–6. *The fundamental equation used in modern wave prediction models. Balances the evolution of the wave spectrum with energy input and dissipation. The same as the* **action balance equation**.

refraction, 50. *The bending of a wave as it propagates over different depths.*

restoring force, 34. *The force that balances that of the wind in wave generation. In other words, that which pulls the sea surface back down again after the wind has lifted it up.*

roaring forties, 20. *A band of westerly winds that blow around the globe at a latitude of approximately 40° south. The band is wider and the winds are stronger in winter.*

rule of twelfths, 108. *Quick method of predicting the tide, used by yachtsmen. Assumes the tide moves a progressive number of twelfths of its full range, every hour.*

S

sea breeze, 12, 86. *A local, onshore wind that blows at certain times of the day in warm climates. Depends on temperature differences between land and sea.*

selective refraction, 82. *The bending of waves of different wavelengths by different amounts when they propagate over changing depth contours.*

self-organizing system, 73. *A system which, through a set of feedback loops and convergent chaotic pathways, organizes itself into a regular pattern.*

semi-diurnal, 103. *Twice a day.*

settiness, 46. *The characteristics of the waves in terms of how they are grouped – e.g., the number, and variability of the number, of waves in the set, and the time between the arrival of successive sets. Same as* **groupiness**.

set-up, 109. *When there are large waves or a strong onshore wind, water is literally pushed up on to the shore, causing the average water surface to slope upwards towards the land.*

significant wave height, 128. *A wave height originally defined to be approximately equal to the most common height estimated by an experienced observer. Significant wave height is normally quantified as the average height of the highest third of all waves observed.*

SMB theory, 79, 85. *An* **empirical** *method of wave prediction invented by H. Sverdrup, W.*

Munk and C. Bretschneider in the 1940s.

Snell's law, 59. *Law governing refraction. Can be used to relate the angle a wave bends to the depths of water over which it propagates. Named after Willebrord van Roijen Snell (1591–1626).*

specific heat capacity, 19. *The amount of heat energy required to raise a specific mass of a substance a certain temperature. In SI units, this would be the amount of heat required to raise 1 kilogram 1 degree Celsius.*

spectrum, 38. *The amount of wave energy at each different period. Can be displayed as a two-dimensional plot.*

spilling breaker, 64. *Technical term for a mushy, slow-breaking wave with no tube.*

spring-neap cycle, 104. *The fortnightly cycle of alternate large and small tidal ranges.*

statistical highs and lows, 16. *Imaginary high and low pressure systems derived from the average pressure over a given time, usually the whole summer or whole winter.*

Stokes drift, 41. *Wave-induced current or wave drift. In shallow water, the orbital motions of the water particles beneath the waves are not exactly circular. They have a forward spiralling motion, resulting in a slow forward movement of water.*

Stokes wave theory, 61. *A slightly more complex set of equations than* **Airy wave theory**, *used to describe wave*

motion where the depth of water is significant compared with the wave height. Named after George Gabriel Stokes (1819–1903).

storm surge, 109. *A sudden increase in the water level at the coast due to a large swell or onshore wind. Strictly speaking, the difference between the observed and predicted sea level. Can have disastrous consequences.*

Storm-Tide Warning Service, 109.

summer profile, 71. *Typical shape of a beach after a long period of small waves – a shoreline* **berm** *and no offshore bar.*

surging breaker, 64. *A wave found on steep shores. One that does not break as such, but rather just 'washes' up and down the beach.*

swell front, 124. *On some wave prediction charts, a line marking a discontinuity in wave period, signifying the arrival of longperiod waves at the beginning of a swell.*

syzygy, 104. *The astronomical configuration for producing spring tides – i.e., opposition or conjunction.*

T

threshold velocity, 70. *In the study of sediment transport, the minimum water velocity required to start moving grains of sediment on the sea-floor.*

tidal curve, 108. *Graph showing the tidal height plotted against time for a particular coastal location.*

tidal range, 73. *The difference between the water level at low tide and the water level at high tide.*

tropical revolving storm (TRS), 118. *This is small, intense area of low pressure that forms between about 5° and 20° latitude, and can be useful for producing surf at certain times of the year. Also called a* **hurricane**, **cyclone** *or* **typhoon**, *depending on the area.*

turbulent eddies, 34. *Small, unpredictable, swirling vortices in a fluid.*

typhoon, 118. *The name given to a* **tropical revolving storm** *(TRS) in the Western Pacific, particularly around Japan and the Philippines.*

U

upwelling, 13. *A phenomenon seen on the western sides of continents, whereby the trade winds constantly blow warm surface water away from the coast, allowing cold water to rise up from underneath, bringing nutrients to the surface and maintaining the surface water cold.*

V

velocity skewness, 71, 76. *When the motion of a fluid is of a higher velocity and shorter duration in one direction, and of a lower velocity and longer duration in the other direction.*

W

WAM, 32. *Stands for WAve Model. Usually refers to the third generation wave prediction model. A large computer program that is used to predict waves all over the oceans.*

warm front, 27. *Imaginary line marking the transition between cold and warm air in a low pressure system. Signified on the weather chart by a thick line with round blobs on it.*

warm sector, 27. *Between the warm and cold fronts in a low pressure system. This is where the wind normally blows strongest in a straight line.*

wave-current interaction, 119. *Effect of current on waves. For example, a current flowing against incoming waves will slow down the waves, causing the wave front to bend (wave-current refraction), and the waves to steepen (increase in height and decrease in wavelength).*

wave frequency, 126. *The inverse of wave period in units of Hertz (Hz), or cycles per second.*

wave period, 10. *The time taken between the passing of one wave crest and the next.*

wave steepness, 65. *The height of a wave divided by its wavelength.*

wave time-series, 128. *A measure of the time history of the sea surface, often obtained from a wave buoy, from which wave height and wave period statistics can be extracted.*

wavelength, 43. *The horizontal distance between one wave crest and the next.*

whitecapping, 35. *When the wind is strong enough to force the tops of existing waves to break, thus removing energy from the waves.*

windsea, 11. *A mixed-up sea containing waves of many different heights, wavelengths and directions. In a windsea, waves are still being generated by the wind, and have not started propagating away on their own.*

winter profile, 71. *Typical shape of a beach after a storm or a period of large waves. The sand has been taken from the shoreline and deposited on an offshore bar.*

ACKNOWLEDGEMENTS

The photographs in this book were supplied by Phil Holden (www.surfsup-mag.co.uk), except for the following: Pete Frieden, p. 56; Alison Hodge, p. 16; Joli, pp. 51, 91 (top), 92; Christopher Laughton, pp. 135, 137; 2002 Orbital Imaging Corporation/Science Photo Library, p. 25; Bernie Robinson, p. 69; Mike Searle, pp. 62, 63, 65; Ester Spears, p. 55. The cover photograph was supplied by Robert Gilley. The authors and publishers are most grateful for permission to reproduce these photographs in *Surf Science*.

Weather charts are reproduced by permission of: European Centre for Medium-Range Weather Forecasts: p. 30 (3.7); Fleet Numerical Meteorology and Oceanography Center: pp. 37 (4.4), 97 (11.2); Instituto National de Meteorologia, Spain: p. 126 (14.6); Meteorological Office, UK: pp. 29 (3.4–3.6), 122 (14.1), 123 (14.2); NOAA/NWS/NCEP Environmental Modeling Center, Ocean Modeling Branch: pp. 124 (14.4), 125 (14.5), 127 (14.7).